Managing Innovative Projects and Programs

It has been estimated that over 75% of the innovative projects that begin through the Innovation Management System (IMS) are either failures or they failed to produce the desired results. The biggest wastes most medium- to large-size organizations face are the waste of money, time, reputation, opportunity, and income that these failures are costing them. Following this book's recommendations could reduce this failure rate by as much as 70%. The purpose of this book is to provide a step-by-step procedure on how to process a medium- or large-size project, program, or product using an already-established IMS that considers the guidance given in ISO 56002:2019 – Innovation Management Systems Standard. Often the most complicated, complex, difficult, and challenging system used in an organization is the IMS. At the same time, it usually is the most important system because it is the one that generates most of the value-adding products for the organization, and it involves all of the key functions within the organization. The opportunity for failure in time and the impact on the organization is critical and often means the difference between success and bankruptcy. Throughout this book, the authors detail the high-impact inputs and activities that are required to process individual projects/programs/products through the innovation cycle. Although this book was prepared to address how medium to large projects, programs, and products proceed through the cycle, it also provides the framework that can be used for small organizations and simple innovation activities. Basically, the major difference between large- and small-impact innovation projects is that the small projects can accept more risks and require fewer resources to be committed. It's important to remember that the authors are addressing an existing IMS rather than trying to create an entirely new one. Currently, this is the only book geared for professionals responsible for managing innovative projects and programs using ISO 56002:2019 – Innovation Management – Innovation Management System – Guidance to provide a comprehensive management strategy and step-by-step plan. It provides a comprehensive analysis of what is required from the time an opportunity is recognized to the time the customer is using the innovative product.

The Management Handbooks for Results Series

The Organizational Alignment Handbook
A Catalyst for Performance Acceleration
H. James Harrington & Frank Voehl (2011)

The Organizational Master Plan Handbook
A Catalyst for Performance Planning and Results
H. James Harrington & Frank Voehl (2012)

The Lean Six Sigma Black Belt Handbook
Tools and Methods for Process Acceleration
H. James Harrington, Frank Voehl,
Chuck Mignosa & Rich Charron (2013)

The Lean Management Systems Handbook
H. James Harrington, Frank Voehl,
Hal Wiggin & Rich Charron (2014)

Change Management
Manage the Change or It Will Manage You
H. James Harrington & Frank Voehl (2016)

Lean Triz
How to Dramatically Reduce Product-Development Costs
with This Innovative Problem-Solving Tool
H. James Harrington (2017)

Innovative Change Management (ICM)
Preparing Your Organization for the New Innovative Culture
H. James Harrington (2018)

The Framework for Innovation
A Guide to the Body of Innovation Knowledge
H. James Harrington, Frank Voehl,
Rick Fernandez & Brett Trusko (2018)

Total Innovative Management Excellence (TIME)
The Future of Innovation
H. James Harrington & Frank Voehl (2020)

Using the ISO 56002 Innovation Management System
A Practical Guide for Implementation and Building a Culture of Innovation
Sid Ahmed Benraouane & H. James Harrington (2021)

Managing Innovative Projects and Programs
Using the ISO 56000 Standards for Guidance and Implementation
H. James Harrington & Sid Ahmed Benraouane (2022)

Managing Innovative Projects and Programs

Using the ISO 56000 Standards for Guidance and Implementation

H. James Harrington
Sid Ahmed Benraouane

A PRODUCTIVITY PRESS BOOK

First Published 2023
by Routledge
605 Third Avenue, New York, NY 10158

and by Routledge
4 Park Square, Milton Park, Abingdon, Oxon, OX14 4RN

Routledge is an imprint of the Taylor & Francis Group, an informa business

ISBN: 978-1-032-19762-3 (hbk)
ISBN: 978-1-032-30601-8 (pbk)
ISBN: 978-1-003-30586-6 (ebk)

DOI: 10.4324/b22993

Typeset in Minion
by Deanta Global Publishing Services, Chennai, India

This book is dedicated to Prof. Mohamed Zairi, a thought leader on quality in the world and a guru of business excellence in Europe and the Middle East and North Africa region. His contribution to the Excellence movement within the United Arab Emirates, and particularly within Dubai, has had a very significant impact and has transformed Dubai into a leading benchmark in the world on excellence, innovation, and customer service. He was a brilliant individual who was always willing to share more than he got back. The world is going to miss this great man and we are going to miss his true and loving friendship.

H. James Harrington and Sid Ahmed Benraouane

Contents

Acknowledgments xv
Preface xvii
About the Authors xxv

Chapter 1 Introduction to Project Innovation Cycle 1

Introduction to Project Innovation 1
Today's Dilemma 3
Addressing the "Unaddressable" 4
Factors Affecting Innovation 7
The Five Types of Innovation 8
The Three Subcategories of Innovation 8
Conclusions Related to Types of Innovations 11
Common Creativity Innovation Killers 11
Your Creative and Innovative Powers 13
The 10 Ss 14
Clauses 1, 2, and 3 15
Knowledge Management System 16

Chapter 2 ISO 56002:2019 19

Introduction 19
TIME 20
ISO 56000:2020 Standards 21
ISO 56002:2019 Standard 22
Clause 4.0 – Context of the Organization 24
Clause 5.0 – Leadership 25
Clause 6.0 – Planning 26
Clause 7.0 – Support 27
Clause 8.0 – Operations 28
Clause 9.0 – Performance Evaluation 28
Clause 10.0 – Improvement 29
More Detailed View of Section 8: Operational 30
Clause 8.1: Operational Planning and Control 30

Collaboration and Partnership 31
Clause 8.2: Managing Innovation Initiatives....................36
Subclause 8.2.1: Managing Each Innovative Initiatives..37
Subclause 8.2.2: Identify How to Implement Each Innovative Initiative 39
Clause 8.3: Innovation Processes.......................................41
Subclause 8.3.1: IMS Overview 41
Building Blocks of the TIME Pyramid.................................. 43
The Advantages of a Good IMS 51
A Word of Caution 53
IMS Summary..53

Chapter 3 Assessing the IMS: A Discussion of 56004:2019 55

Assessment Approaches..55
Assessment Objectives 56
The Breath and the Extent of the Assessment...................57
Assessment Focus 57
Expertise Involved in Assessment.......................................58
Data Collection and Data Collection Tools 58
Data Type 59
Reference and Comparison 59
Data Interpretation 60
Innovation Management Outputs, Formats, and Reports 60
Section 5.2.1: Performance Criteria for Innovation Management..62

Chapter 4 The PIC Cast Members 65

Introduction 65
The Cast of Players...65
Research and Development 70
Product Engineering 72
Manufacturing Engineering 73
Marketing..74
Executive Project Sponsor 76

Project Team Leaders 77
Team Leaders....................78
Project Managers 78
Innovation Project Team 83
Team Ground Rules.................... 84

Chapter 5 The PIC Overview 89

ISO 56002:2019 Innovation Processes Cycle 89
PIC Phases 90
PIC Tollgates....................93
Five Tollgates Summary.................... 96
Introduction to PIC Three Phases 97
Six Levels of Documentation 99
How to Apply the Six Levels of Documentation to Innovation....................102
Level 0 – Organizational....................102
Level I – Systems 102
Level II – Phase Levels Are Divided into Process Groupings....................103
Level III – Process Groupings Are Divided into Processes....................104
Level IV – Process Levels Are Divided into Activities 105
Level V – Activities Are Divided into Tasks 107
Activity Block Diagrams 108

Chapter 6 Phase I: Creation 111

Introduction 111
Process Grouping 1: Opportunity Identification 111
Inputs to Process Grouping 1: Opportunity Identification 115
Process Grouping 1: Opportunity Identification Activity Block Diagram 119
Tollgate I.................... 123
Process Grouping 2: Opportunity Development 125
Objective of Process Grouping 2: Opportunity Development.................... 126

Process Grouping 2: Opportunity Development Activity Block Diagram for Apparent and/or Minor Opportunities....129
Process Grouping 2: Opportunity Development Activity Block Diagram for Major Opportunities....130
Process Grouping 2: Opportunity Development Activity Block Diagram for New Paradigms and Discovery Opportunities....132
Process Grouping 3: Value Proposition....135
Inputs to Process Grouping 3: Value Proposition....136
Process Grouping 3: Value Proposition and Tollgate II Activity Block Diagram....142
Tollgate II: Concept Approval....146
Process Grouping 4: Concept Validation....147
Process Grouping 4: Concept Validation Activity Block Diagram....147
Summary of Phase I: Creation....150

Chapter 7 Phase II: Preparation and Producing....151

Introduction....151
Tollgate III: Project Approval....153
Tollgate III: Activity Block Diagram....154
Tollgate III: Project Approval Supports the Business Case Analysis....158
Process Grouping 5: Business Case Analysis....159
Inputs to Process Grouping 5: Business Case Analysis....160
Business Case Validation....162
Document Performance and Project Resource Requirements for Each Project/Program....162
Business Cases That Do Not Require Additional Resources....162
Criteria to Rank Business Cases....163
Results of Business Case Approval....164
Process Grouping 5: Business Case Analysis Activity Block Diagram....165
Summary of Process Grouping 5: Business Case Analysis and Tollgate III....169

Process Grouping 6: Resource Management........................169
Inputs to Process Grouping 6: Resource Management........170
Process Grouping 6: Human Resource Staffing Activity Block Diagram...171
Process Grouping 6: Facilities Resource Management Activity Block Diagram.............................174
Floor Space Resource Management Activity Block Diagram...174
Equipment Resource Management Activity Block Diagram...176
Process Grouping 6: Facilities Setup Resource Management Activity Block Diagram.............................177
Process Grouping 6: Financial Resources Management Activity Block Diagram.............................179
Process Grouping 7: Documentation.....................................181
Six Document Management Systems.............................182
Process Grouping 7: Part 1 – Product Specifications Documentation Activity Block Diagram 183
Process Grouping 7: Part 2 – Project Management Plan and Facilities Planning Documentation Activity Block Diagram 185
Process Grouping 7: Part 3 – Suppliers and Contractors Documentation Activity Block Diagram...... 187
Process Grouping 7: Part 4 – Production Setup Documentation Activity Block Diagram 188
Process Grouping 7: Part 5 – Producing Output Controls Documentation Activity Block Diagram........ 190
Process Grouping 7: Part 6 – Marketing and Sales Documentation Activity Block Diagram 192
Summary Process Grouping 7: Documentation 195
Process Grouping 8: Production...195
Process Grouping 8: Production Activity Block Diagram 196
Tollgate IV: Customer Ship Approval....................................199
Tollgate IV: Customer Ship Approval Activity Block Diagram 200
Summary Process Grouping 8: Production 203

Chapter 8 Phase III: Delivery 207
Introduction 207
Process Grouping 9: Marketing, Sales, and Delivery 207
Process Grouping 9: Marketing, Sales, and Delivery Activity Block Diagram 211
Summary of Process Grouping 9: Marketing, Sales, and Delivery 214
Process Grouping 10: After-Sales Service Activities 215
Process Grouping 10: After-Sales Services Activity Block Diagram 217
Process Grouping 11: Performance Analysis 220
Tollgate V: Project Evaluation 221
Typical Improvement Methodology Results 221
Top Five Positive/Negative Innovation Change Impacts 222
Process Grouping 11: Performance Analysis Activity Block Diagram 225
Summary 228
Process Grouping 12: Transformation 229
Process Grouping 12: Transformation Activity Block Diagram 231
Summary 235

Chapter 9 Special Mention Tools and Methodologies 237
Introduction 237
Area Activity Analysis: A Key 2020s Tool Put on the Back Burner 238
Introduction 238
Benefits of AAA 240
Features of AAA 241
Background 241
AAA Methodology 242
Phase I: Preparation for AAA 243
Phase II: Develop Area Mission Statement 243
Phase III: Define Area Activities 245
Phase IV: Develop Customer Relationships 245
Phase V: Analyze the Activity's Efficiency 246

Phase VI: Develop Supplier Partnerships 246
Phase VII: Performance Improvement 247
Summary 248
Benchmarking for Better Innovation 249
Introduction 249
Benchmarking Cycle Time 254
Evaluating Competitive Products 254
Benchmarking Code of Conduct 255
Benchmarking Protocol 258
Examples 259
Software 261
Design for X 261
Introduction 261
User 264
How to Use the Tool 265
Knowledge Management System 266
Design Guidelines 266
Design Analysis Tools 267
Design for X Procedure 267
Design for Safety 269
Design for Reliability 270
Design for Testability 270
Design for Assembly/Manufacturing 271
Design for Environment 271
Design for Serviceability 272
Design for Ergonomics 272
Design for Aesthetics 272
Design for Packaging 273
Design for Features 273
Design for Time to Market 273
Summary 274
Knowledge Management Excellence 274
What Is Knowledge? 275
Implementing a KMS 276
Starting and Implementing a KMS 276
KMS Maturity Grid 279
Risk Management 280
Introduction 280

Risk Management Framework....281
A Typology of Risks....281
Ways of Dealing with Risk....282
Risk Management Framework: ISO 3100 Guidelines....283
The Risk Assessment Process....285
Risk Management in ISO 56002:2019....288
Surveys....289
Designing Survey Questions....291
Notes....293

Bibliography....295

Appendix A: The *Enhance Solutions Platform*™ for Innovation Management....297

Appendix B: Most Used Tools....305

Index....309

Acknowledgments

We would like to first acknowledge all of the individuals who participated in ISO Technical Committee 279 for their willingness to share their knowledge, ideas, and thoughts with the rest of the team to create an international agreement related to innovation concepts. Without that free exchange of knowledge, the innovation ISO standards would not have been able to be created.

Secondly, we would like to acknowledge the hard work and dedication of Candy Rogers and Michael Sinocchi who have made it possible for translating a rough draft manuscript into a finished book that we both are proud of. Last, but not least, we would like to acknowledge and give deserving special recognition to the International Organization for Standardization (ISO) and ANSI for their contributions in organizing and contributing to the management of Technical Committee 279.

Preface

INTRODUCTION

"A Book for Managers of Innovative Projects/Programs"

It has been estimated that over 75% of the innovative projects that start through an Innovation Management System (IMS) are failures or failed to produce the desired results. One of the biggest wastes most medium-to-large-size organizations are facing is the waste of money, time, reputation, opportunity, and income that these failures are costing them. Following this book's recommendations could reduce this failure rate by as much as 70%.

The purpose of this book is to provide a step-by-step procedure on how to process a medium- or large-size project/program/product using an already-established Innovation Management System (IMS) where we take into consideration the guidance given in ISO 56002:2019 – Innovation Management Systems Standard.

Often the most complicated, complex, difficult, and challenging system used in an organization is the IMS. At the same time, it usually is the most important system because it is the one that generates most of the value added for the organization and it involves all of the key functions within the organization. The opportunity for failure in time and the impact on the organization is critical and often means the difference between success and bankruptcy. In the following text, we will look at the high-impact inputs and activities that are required to process individual projects/programs/products through the innovation cycle.

Although this book was prepared to address how medium to large projects/programs/products proceed through the cycle, it does provide a framework that can also be used for a smaller organization's simpler innovation activities. Basically, the big difference between large- and small-impact innovation projects is that the small projects can accept more risks and require less resources to be committed. It's important to remember that we are using an existing IMS, rather than trying to improve on an existing one.

The authors have selected the *Enhance Solutions Platform*™ to create a "live extension" to this book. Dynamic process models representing many of the innovation process diagrams in the book have been created using *Enhance* and are available to you at no additional cost through *Enhance's TeamPortal*™ cloud service. *TeamPortal* not only provides a way to dynamically navigate the book's innovation process models but also offers additional knowledge such as documents, book extracts, and links to web-based resources and even videos. Combining this book and *Enhance's TeamPortal*™ cloud service creates a unique and valuable way for you to learn how to implement and manage an innovative project or program within your enterprise. You can access *Enhance* by visiting **EDGESoftware.cloud/managing-innovation.**

The authors do not wish to imply that the *Enhance* cloud service is the only cloud service capable of helping you manage the complex interfaces that occur when an organization's total interconnecting relationships between its individual processes are defined and documented. It is the cloud service that we have developed experience with and feel comfortable in using. There may be other cloud services that perform the functions as well as or even better than *Enhance.* Whatever service you choose to use, we believe that it is necessary to use an automated system if you are going to tie together all the interconnecting points between systems within your organization due to the complexity and amount of overlap and variation.

ISO 56002:2019 – INNOVATION MANAGEMENT SYSTEMS STANDARD

Understanding an IMS Hierarchy

The following are important definitions to understand an IMS typical hierarchy.

- Task – Steps that are required to perform a specific activity
- Activity – Small parts of a process usually performed by a single department or individual

- Process – A series of logically interconnected, related activities where each activity takes inputs, adds value to the activity, and produces an output to an internal or external customer. It's how an organization's day-to-day routine work is accomplished. An organization's processes define how it operates
- Tollgate – A tollgate is the name for a process grouping that contains processes required to evaluate a project/program status and compare the project to a plan before it moves onto the next process grouping
- Project – A project is a temporary endeavor undertaken to create a unique product or service. It has a defined start and end point
- Phase – A distinct period or stage in a series of events or a process of change or development
- Process Grouping – A group of processes that are closely interrelated and required to accomplish a specific objective before the phase to be completed
- System – Groups of related processes that may or may not be connected

In ISO 56002:2019 ANSI presented innovation processes as consisting of the following processes (see Figure P.1):

1. Identify opportunities
2. Create concept
3. Validate concept

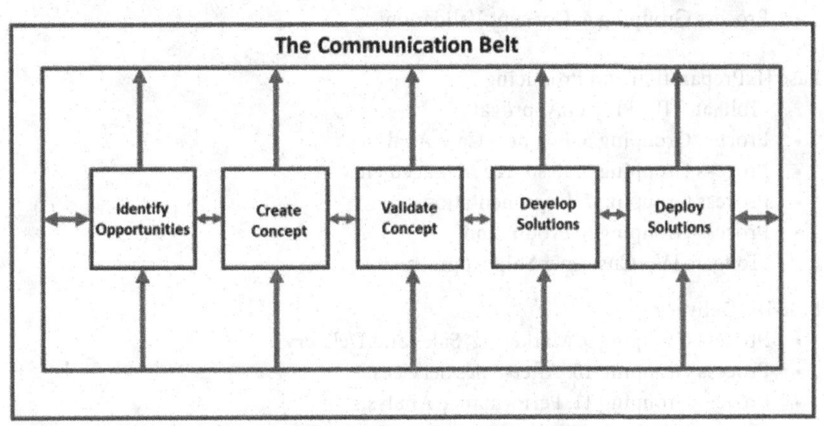

FIGURE P.1
The Five Innovative Processes Linked Together

4. Develop solutions
5. Deploy solutions

As you can see from Figure P.1, there is a great deal of interaction between all five processes. With so many different organizations involved in the system and each of them having their own processes to accomplish their part of the total cycle, the system has to be able to accept a great deal of variation in content and sequencing from project to project. Its primary focus is on designing an innovative product, service, process, etc. that provides value to the stakeholders.

We personally like a three-phase approach that is made up of 12 Process Groupings. We have included two figures for the PIC. The first figure, Figure P.2, is a typical word document. The second figure, Figure P.3, shows the same cycle as a computer would present it. This figure and all the computerized figures in Phase I were created by *Enhance's TeamPortal*™

Project Innovation Cycle (PIC)
12 Innovation Process Groupings

Phase I. Creation

- Process Grouping 1. Opportunity Identification
  *Tollgate I - Opportunity Analysis
- Process Grouping 2. Opportunity Development
- Process Grouping 3. Value Proposition
  *Tollgate II - Concept Approval
- Process Grouping 4. Concept Validation

Phase II. Preparation and Producing
  *Tollgate III - Project Approval

- Process Grouping 5. Business Case Analysis
- Process Grouping 6. Resource Management
- Process Grouping 7. Documentation
- Process Grouping 8. Production
  *Tollgate IV - Customer Ship Approval

Phase III. Delivery

- Process Grouping 9. Marketing, Sales and Delivery
- Process Grouping 10. After-Sales Services
- Process Grouping 11. Performance Analysis
  *Tollgate V - Project Evaluation
- Process Grouping 12. Transformation

FIGURE P.2
PIC Process Groupings and Tollgates

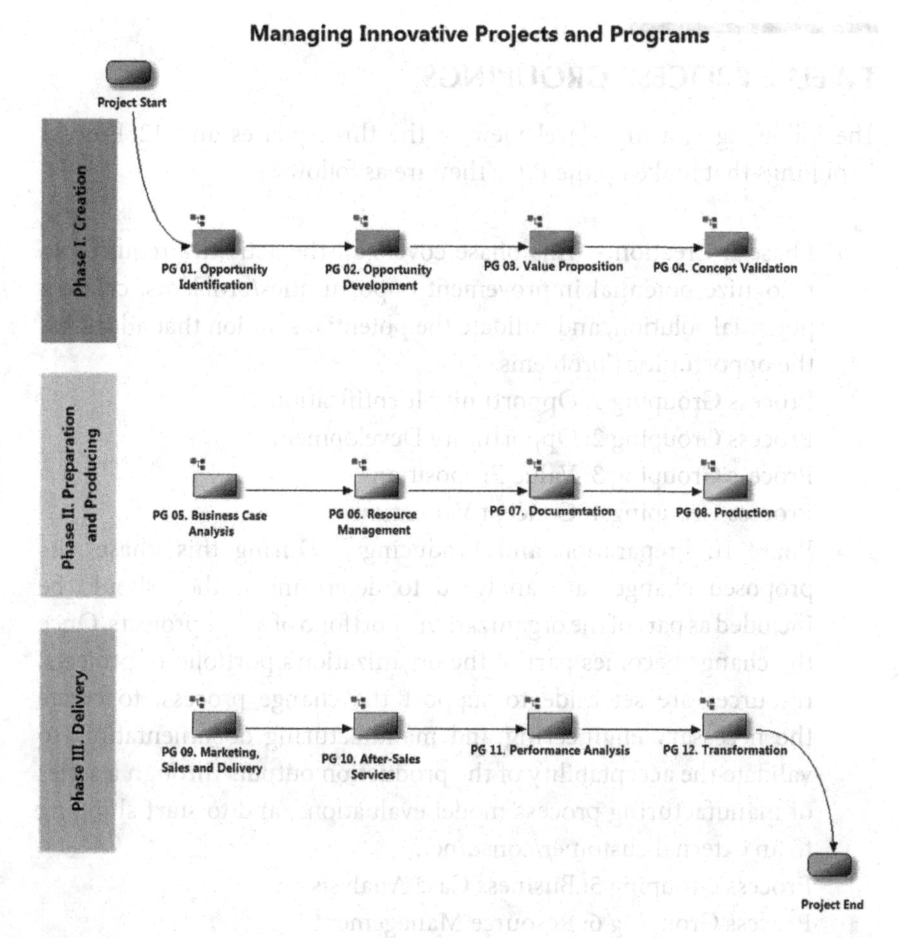

FIGURE P.3
PIC Process Groupings as Seen in EDGE Software's *Enhance TeamPortal* Cloud Service

cloud service, and most are available to you at no additional cost by visiting **EDGESoftware.cloud/managing-innovation.**

This cycle results in an output of an innovative entity that goes to an external customer. To create an innovative entity, much focus must be placed upon the production and selling processes as is placed on the product design processes. An idea or concept may have the potential of being innovative, but it doesn't become reality until it is transformed into a value-added entity.

Our view of the innovative cycle is slightly different from ISO's five-phase approach, although both approaches provide satisfactory innovative results. Which one you use is purely a matter of choice. In this book, we will be using the three-phase approach because it has more details focused on the latter part of the Project Innovation Cycle (PIC).

TWELVE PROCESS GROUPINGS

The following is a high-level view of the three phases and 12 Process Groupings that make up the PIC. They are as follows:

- Phase I: Creation – This phase covers all the activities required to recognize potential improvement opportunities/problems, create a potential solution, and validate the potential solution that addresses the opportunities/problems.
  Process Grouping 1: Opportunity Identification
  Process Grouping 2: Opportunity Development
  Process Grouping 3: Value Proposition
  Process Grouping 4: Concept Validation
- Phase II: Preparation and Producing – During this phase, the proposed changes are analyzed to determine if they should be included as part of the organization's portfolio of active projects. Once the change becomes part of the organization's portfolio of projects, resources are set aside to support the change process, to create the necessary engineering and manufacturing documentation, to validate the acceptability of the production outputs through a series of manufacturing process model evaluations, and to start shipping to an external customer/consumer.
  Process Grouping 5: Business Case Analysis
  Process Grouping 6: Resource Management
  Process Grouping 7: Documentation
  Process Grouping 8: Production
- Phase III: Delivery – During this phase, the output from the process is transformed from items into dollars and cents, profit and losses, success and failures. It also includes a performance analysis to compare actual results to projected value added for each stakeholder.
  Process Grouping 9: Marketing, Sales, and Delivery
  Process Grouping 10: After-Sales Services
  Process Grouping 11: Performance Analysis
  Process Grouping 12: Transformation

Usually the project team is disbanded after the Process Grouping 11 is completed, but that's only the beginning of the project story. The real

FIGURE P.4
Black Beauty Who Has Gone More than 380,000 Miles and Is Still Going Strong

test of the project occurs over the next year or two when the approaches are often reset to the original habit patterns. For successful innovative projects, changes have to become part of the organization's culture and habit patterns. This is where the real impact of the project is evaluated.

We used to be able to win the blue ribbon entitled "Innovator" based on the uniqueness of our product design. Today this is still true, but equally important is personalized and speedy customized customer care and long-term sustained performance that delivers real added value to the stakeholders. My first car was a Model A Ford. It required me to pull the motor out about every 20,000 miles. It had tires that only lasted for 5,000 miles. If you were going on a trip, you had to be smart enough to take along several extra tires. Today I'm driving a Lincoln town car with 380,000 miles on it with no signs of giving me any future problems (Figure P.4, a picture of my black beauty).

SUMMARY

The first objective of this book is to provide you with enough information so that you can decide if you need to improve, redesign, or leave alone your IMS. If we are successful, the next objective is to show you how to manage your individual projects, programs, and portfolios to be able to increase your quantity, creativeness, and frequency of producing new creative products and services that provide you with a competitive advantage. Although this book was prepared to address how medium to large projects/programs/products proceed through the IMS, it does provide a framework that can also be used for small organizations with simpler innovation activities. Basically, the big difference between large- and small-impact innovation projects is that simple, small projects have fewer people involved, are easier to implement, address less resistance, accept more risks, and require less resources. It's important to remember that we are using an existing system rather than trying to improve on an existing IMS. However, using this book often highlights major opportunities to improve the existing IMS.

This book focuses on maximizing the results of the portfolio of projects and programs that the organization is involved in without directly trying to address the cultural and risk strategy that is embedded in the organization. We have observed an excellently designed IMS that has produced unsatisfactory results to the point that it drove the organization into bankruptcy and, on the other hand, organizations with a minimum IMS that created miracles. It's not what you do but how you do it that counts.

Many organizations do a good job. The "great" organizations do it well.

H. James Harrington

About the Authors

H. James Harrington
Chief Executive Officer
Harrington Management Systems

In the book *Tech Trending*, Dr. H. James Harrington was referred to as "the quintessential tech trender." The *New York Times* referred to him as having a "knack for synthesis and an open mind about packaging his knowledge and experience in new ways – characteristics that may matter more as prerequisites for new-economy success than technical wizardry."

It has been said about him, "Harrington writes the books that other consultants use."

The leading Japanese author on quality, Professor Yoshio Kondo, stated: "Business Process Improvement (methodology) investigated and established by Harrington and his group is some of the new strategies which brings revolutionary improvement not only in quality of products and services, but also the business processes which yield the excellent quality of the output."

The father of "Total Quality Control," Dr. Armand V. Feigenbaum, stated: "Harrington is one of those very rare business leaders who combines outstanding inherent ability, effective management skills, broad technology background, and great effectiveness in producing results. His record of accomplishment is a very long, broad and deep one that is highly and favorably recognized."

Author Tom Peters stated, "I fervently hope that Harrington's readers will not only benefit from the thoroughness of his effort but will also 'smell' the fundamental nature of the challenge for change that he mounts."

William Clinton, former President of the United States, appointed Harrington to serve as an Ambassador of Good Will.

Newt Gingrich, former Speaker of the House and General Chairman of American Solutions, has appointed Harrington to the Advisory Board of his Jobs and Prosperity Task Force. Gingrich stated: "Now it's time for the people who know how to create jobs and prosperity, know what it means to stick to a budget and make payroll – the business leaders of the country – to come together and clean up this mess…And we're thrilled to have Harrington on board as we begin the work to get this economy back on track."

In 2009, the Editor of ASQ's official magazine, *Quality Progress*, printed the following statement: "Among those recognized as gurus are W. Edwards Deming, Joseph Juran, Philip Crosby, H. James Harrington, Kaoru Ishikawa, Walter A. Shewhart, Shigeo Shingo, Frederick Taylor, and Genichi Taguchi."

Harrington Management Systems (formerly Harrington Institute) was featured on a half-hour TV program, *Heartbeat of America*, which focuses on outstanding small businesses that make America strong. The host, William Shatner, stated: "You (Dr. Harrington) manage an entrepreneurial company that moves America forward. You are obviously successful."

Wayne Newton's comment related to Harrington was, "Jim, when I grow up, I want to be you."

Key Responsibilities:
Harrington now serves as the Chief Executive Officer (CEO) for the Harrington Management Systems (formerly Harrington Institute) with branches in many countries around the world. Harrington Management Systems is an international consulting system consisting of four divisions – Management Solutions, Technical Solutions, the Harrington Academy, and Certification Activities. At the present time, the Harrington Certification Activities has certified more than 4,000 individuals.

He also serves as

- President of the Walter L. Hurd Foundation;
- Honorary Advisor for Quality for China;
- Chairman of the Centre for Organizational Excellence Research (COER); and
- On the Board of Directors for a number of small- to medium-size companies helping them develop their business strategies.

Harrington is recognized as one of the world leaders in applying performance improvement methodologies to business processes. For 11 years, he wrote a regular column for *Quality Digest Magazine* (U.S.). He also wrote columns for *Business Quality Review* (Dubai) and for several publications in China and Mexico. Harrington is on the Editorial Review Board for five magazines. He has an excellent record of coming into an organization working as its CEO or Chief Operating Officer (COO), resulting in a major improvement in its financial and quality performance.

While Harrington was Chairman of ASQ, he was one of the leaders in getting the Malcolm Baldrige National Quality Award through Congress and approved by the President of the United States. He also served as the first Treasurer of the Malcolm Baldrige Consortium that set up and developed the award.

Previous Experience:

From 2015 to 2017, Harrington served as President of Altshuller Institute.

In 2008 and 2009, Harrington served as COO of Define Properties in Dubai, UAE.

From 2004 to 2008, he served as the CEO of Harrington Middle East, a management consulting firm located in the UAE.

In February 2002, he retired as the Chief Operating Officer of Systemcorp A.L.G. Systemcorp, which was later bought by IBM, is the leading supplier of knowledge management and project management solutions for global and national organizations seeking to gain maximum financial value from the alignment of IT and business projects with their organization's strategies. As COO of Systemcorp, his primary objective was to install effective processes that will support the company going public. While he was with Systemcorp, they completed one round of pre-IPO funding. Harrington's focus was directed at establishing a high-quality professional service group, an international sales team, and a controlled product engineering organization. Under his leadership, product sales increased by 200%. He actively pursued alliance partnerships to serve as sale channels. Agreements were signed with IBM, Standards & Poors, Albertsons, and Abbott Laboratories. In December 2000, IBM, Standards & Poors, Albertsons, and Abbott Laboratories bought a share in the company. IBM also became part owner of Systemcorp and signed a Strategic Alliance Partnership agreement with Systemcorp. IBM also established a business unit around Systemcorp's product, PMOffice.

Prior to becoming COO at Systemcorp, Harrington had joined Ernst & Young in October 1989 when Ernst & Young acquired his company Harrington, Hurd, & Rieker (HH&R). For the next 10 years, he served as a Principal and one of the leaders in the Performance Improvement Group at Ernst & Young. In this assignment, he helped Ernst & Young develop its methodology and provided managerial and technical direction to project teams. He also served as their International Quality Advisor. During this period of time, he worked as the leader of projects at companies like:

- U.S. Army installations at Cape Kennedy
- McDonnell Douglas
- Kraft
- Labatt's
- General Dynamics

Harrington helped Ernst & Young develop its methodologies for Process Reengineering, Knowledge Management, ISO 9000, and Change Management. He left Ernst & Young in January 2000 to join Systemcorp.

In 1987, he left IBM to start a consulting firm called Harrington and Hurd Associates. In the latter part of 1987, he acquired Rieker Management Systems, forming a new company called Harrington, Hurd & Rieker, Inc. (HH&R). He served as its President and CEO until it was acquired by Ernst & Young. HH&R provided performance improvement training and consulting services. It was the first firm to define a methodology for "Process Redesign," which was laid out in Harrington's best-selling book *Business Process Improvement*. Under his creative leadership, the business expanded rapidly, becoming profitable in the first year and growing at more than 80% per year that led to Ernst & Young buying the organization.

HH&R specialized in:

- Process Redesign
- TQM
- Team Concepts
- Design of Experiments
- Knowledge Management
- Supply Chain Management
- Strategic Planning
- Improvement Planning

Prior to starting HH&R, Harrington worked at IBM for 40 years. He started as an apprentice toolmaker and rose to the level of Senior Engineer and Project Manager. During his time with IBM, he gained a wealth of practical experience in serving in Executive Management positions in manufacturing, test engineering, product engineering, reliability engineering, and quality assurance. While he was at IBM, he had assignments in the United States, Japan, Singapore, Germany, and Great Britain. He served as a Project Manager for all of the random access files and supporting software that were developed in the General Products Division of IBM between 1975 and 1985.

Some of his major accomplishments while at IBM include:

- Established and led IBM's Quality Research Center
- Designed their Process Qualification procedures
- Designed their Process Capability (Benchmarking) procedure
- Established Office of Line Product Reliability Program
- Part of the team that developed the Total Quality Analysis and Report Software System
- Installed Poor-Quality Cost
- Established a Failure Analysis Laboratory
- Developed the approach to integrating quality into the total organization
- Part of the team that established IBM's Quality Education System
- Part of the team that prepared the Corporate Quality Manual

Harrington is past Chairman and past President of the prestigious International Academy for Quality and of the American Society for Quality Control. He is also an active member of the Global Knowledge Economics Council.

He served for 10 years as an A-level member of ISO TC176 (the technical committee responsible for writing the ISO 9000 series). He also served as an A-level member of TC 207 (the technical committee responsible for writing ISO 14000 environmental standards) for five years representing the International Academy for Quality. He is now a member of the ISO TC 279 that is responsible for preparing the standard for Innovation Management Systems.

He was honorably discharged from the U.S. Navy after four years of service in May 1955. His service number (military identification number) was 785–92–27.

Credentials:
Harrington was elected to the honorary level of the International Academy for Quality, which is the highest level of recognition in the quality profession.

Harrington is a government-registered Professional Quality Engineer, a Certified Quality Engineer, and a Certified Reliability Engineer by the American Society for Quality Control, and a Permanent Certified Professional Manager by the Institute of Certified Professional Managers. He is a certified Master of Six Sigma Black Belt and received the title of Six Sigma Grand Master. Harrington has an MBA and PhD in Engineering Management and a BS in Electrical Engineering. In 2013, a degree of Doctor of Philosophy was conferred upon Harrington by the Sudan Academy of Sciences for his "immense contributions, remarkable achievements and distinguished accomplishments in the field of Quality Management, Business Excellence, and Innovation, covering wider range of geographical locations and countries."

Harrington has served as the Chairman of the Advisory Board for E-TQM College and as a Member of the Hamdan Bin Mohammed e-University Advisory Board. Now he is an Honorary Advisory Board member for the Hamdan Bin Mohammed e-University Advisory Board.

Harrington's contributions to performance improvement around the world have brought him many honors. He was appointed the honorary advisor to the China Quality Control Association and was elected to the Singapore Productivity Hall of Fame in 1990. He has been named Lifetime Honorary President of the Asia-Pacific Quality Control Organization and Honorary Director of the Association Chilean de Control de Calidad. In 2006, he accepted the Honorary Chairman position of Quality Technology Park of Iran.

Harrington has been elected a Fellow of the British Quality Control Organization and the American Society for Quality Control. In 2008, he was elected to be an Honorary Fellow of the Iran Quality Association and of the Azerbaijan Quality Association. He was also elected an honorary member of the quality societies in Taiwan, Argentina, Brazil, Colombia, and Singapore. He is also listed in the "Who's-Who Worldwide" and "Men of Distinction Worldwide." He has presented hundreds of papers on performance improvement and organizational management structure at the local, state, national, and international levels.

Recognition/Medals/Awards:

- The Harrington/Ishikawa Medal, presented yearly by the Asian Pacific Quality Organization, was named after Harrington to recognize his many contributions to the region.
- The Harrington/Neron Medal was named after Harrington in 1997 for his many contributions to the quality movement in Canada.
- Harrington Best TQM Thesis Award was established in 2004 and named after Harrington by the European Universities Network and e-TQM College.
- Harrington Chair in Performance Excellence was established in 2005 at the Sudan University.
- Harrington Excellence Medal was established in 2007 to recognize an individual who uses the quality tools in a superior manner.
- H. James Harrington Scholarship was established in 2011 by the ASQ Inspection Division.
- Harrington Center for Quality and Excellence was established in 2014 in Sudan in recognition of all his contributions all over the world.

Harrington got his first award in 1960 when he received the Mac Titan Award from NASA for improving yield on the Titan missile project at IBM. Over the past 50 years, he has received many awards, including the Benjamin L. Lubelsky Award, the John Delbert Award, the Administrative Applications Division Silver Anniversary Award, and the Inspection Division Gold Medal Award.

Additional Awards Harrington Has Received:
In 1989, Harrington was appointed as an honorary advisor of the third Board of Directors of China Quality Control Association. In 1996, he received the ASQC's Lancaster Award in recognition of his international activities. In 1999, he was appointed as an advisor of the Shanghai Academy of Quality Management. In 2001, he received the Magnolia Award in recognition for the many contributions he has made in improving quality in China. In 2002, Harrington was selected by the European Literati Club to receive a lifetime achievement award at the Literati Award for Excellence ceremony in London. The award was given to honor his excellent literature contributions to the advancement of quality and organizational

performance. Also, in 2002, Harrington was awarded the International Academy of Quality President's Award in recognition for outstanding global leadership in quality and competitiveness, and contributions to IAQ as Nominations Committee chair, vice president, and chairman. In 2003, Harrington received the Edwards Medal from the American Society for Quality (ASQ). The Edwards Medal is presented to the individual who has demonstrated the most outstanding leadership in the application of modern quality control methods, especially through the organization and administration of such work. In 2004, he received the Distinguished Service Award, which is ASQ's highest award for service granted by the Society. In 2008, Harrington was awarded the Sheikh Khalifa Excellence Award (UAE) in recognition of his superior performance as an original Quality and Excellence Guru who helped shape modern quality thinking. In 2009, Harrington was selected as the Professional of the Year (2009). Also in 2009, he received the Hamdan Bin Mohammed e-University Medal. In 2010, the Asian Pacific Quality Association (APQO) awarded Harrington the APQO President's Award for his "exemplary leadership." The Australian Organization of Quality NSW's Board recognized Harrington as the Global Leader in Performance Improvement Initiatives in 2010. In 2011, he was honored to receive the Shanghai Magnolia Special Contributions Award from the Shanghai Association for Quality in recognition of his 25 years of contributing to the advancement of quality in China. This was the first time that this award was given out. In 2012, Harrington received the ASQ Ishikawa Medal for his many contributions in promoting the understanding of process improvement and employee involvement on the human aspects of quality at the local, national, and international levels. Also in 2012, he received the Jack Grayson Award, which recognizes individuals who have demonstrated outstanding leadership in the application of quality philosophy, methods, and tools in education, healthcare, public service, and not-for-profit organizations. Harrington also received the A.C. Rosander Award in 2012. This is ASQ Service Quality Division's highest honor. It is given in recognition of outstanding long-term service and leadership resulting in substantial progress toward the fulfillment of the Division's programs and goals. Additionally, in 2012, Harrington was honored by the Asia Pacific Quality Organization by being awarded the Armand V. Feigenbaum Lifetime Achievement Medal. This award is given annually to an individual whose relentless pursuit of performance improvement over a minimum of 25 years has distinguished himself or herself for the candidate's work in promoting the use of quality methodologies and principles within and outside of the organization he or she is part of. In 2017, he received the first Lifetime Achievement Award for his many contributions to the advancement of process improvement and innovation methodologies from the Innovation Association of Innovative Professionals (IAOIP).

Books by Harrington:
Harrington is a very prolific author, publishing hundreds of technical reports and magazine articles. He has authored or coauthored over 55 books, which are:

- *The Improvement Process*; 1987 – one of 1987 best-selling business books
- *Poor-Quality Cost*; 1987
- *Excellence – The IBM Way*; 1988
- *The Quality/Profit Connection*; 1988
- *Business Process Improvement*; 1991 – the first book on Process Redesign
- *The Mouse Story*; 1991
- *Of Tails and Teams*; 1994
- *Total Improvement Management*; 1995
- *High Performance Benchmarking*; 1996
- *The Complete Benchmarking Workbook*; 1996
- *ISO 9000 and Beyond*; 1996
- *The Business Process Improvement Workbook*; 1997
- *The Creativity Toolkit – Provoking Creativity in Individuals and Organizations*; 1998
- *Statistical Analysis Simplified – The Easy-to-Understand Guide to SPC and Data Analysis*; 1998
- *Area Activity Analysis – Aligning Work Activities and Measurements to Enhance Business Performance*; 1998
- *ISO 9000 Quality Management System Design: Optimal Design Rules for Documentation, Implementation, and System Effectiveness (ISO 9000 Quality Management System Design)* – co-author; 1998
- *Reliability Simplified – Going beyond Quality to Keep Customers for Life*; 1999
- *ISO 14000 Implementation – Upgrading Your EMS Effectively*; 1999

- *Performance Improvement Methods – Fighting the War on Waste*; 1999
- *Simulation Modeling Methods – An Interactive Guide to Results-Based Decision Making*; 2000
- *Project Change Management – Applying Change Management to Improvement Projects*; 2000
- *E-Business Project Manager*; 2002
- *Process Management Excellence – The Art of Excelling in Process Management*; 2005
- *Project Management Excellence – The Art of Excelling in Project Management*; 2005
- *Change Management Excellence – The Art of Excelling in Change Management*; 2005
- *Knowledge Management Excellence – The Art of Excelling in Knowledge Management*; 2005
- *Resource Management Excellence – The Art of Excelling in Resource Management*; 2005
- *Six Sigma Statistics Simplified*; 2006
- *Improving Healthcare Quality and Cost with Six Sigma; 2006*
- *Making Teams Hum*; 2007
- *Advanced Performance Improvement Approaches: Waging the War on Waste II*; 2007
- *Six Sigma Green Belt Workbook*; 2008
- *Six Sigma Yellow Belt Workbook*; 2008
- *(FAST) Fast Action Solution Teams*; 2008
- *Strategic Performance Improvement Approaches: Waging the War on Waste III*; 2008
- *Corporate Governance: From Small to Mid-Sized Organizations*; 2009
- *Streamlined Process Improvement*; 2011
- *The Organizational Alignment Handbook: A Catalyst for Performance Acceleration*; 2011
- *The Organizational Master Plan Handbook: A Catalyst for Performance Planning and Results*; 2012
- *Performance Accelerated Management (PAM): Rapid Improvement to Your Key Performance Drivers*; 2013
- *Closing the Communication Gap: An Effective Method for Achieving Desired Results*; 2013
- *Lean Six Sigma Black Belt Handbook: Tools and Methods for Process Acceleration*; 2013

- *Lean Management Systems Handbook*; 2014
- *Maximizing Value Propositions to Increase Project Success Rates*; 2014
- *Making the Case for Change: Using Effective Business Cases to Minimize Project and Innovation Failures*; 2014
- *Techniques and Sample Outputs that Drive Business Excellence*; 2015
- *Effective Portfolio Management Systems*; 2015
- *Change Management: Manage the Change or It Will Manage You*; 2015
- *The Innovation Tools Handbook, Volume 1: Organizational and Operational Tools, Methods, and Techniques That Every Innovator Must Know*; 2016
- *The Innovation Tools Handbook, Volume 2: Evolutionary and Improvement Tools That Every Innovator Must Know*; 2016
- *The Innovation Tools Handbook, Volume 3: Creative Tools, Methods, and Techniques That Every Innovator Must Know*; 2016
- *Lean TRIZ: How to Dramatically Reduce Product-Development Costs with This Innovative Problem-Solving Tool*; 2017
- *The Framework for Innovation: A Guide to the Body of Innovation Knowledge*; 2018
- *Project Management for Performance Improvement Teams*; 2018
- *Creativity, Innovation, and Entrepreneurship*; 2018
- *Innovative Change Management*; 2018
- *The Innovation Systems Cycle: Simplifying and Incorporating the Guidelines of the ISO 56002 Standard and Best Practices*; 2019
- *Total Innovation Management Excellence (TIME)*; May 2020
- *Structuring Your Organization for Innovation*; 2020
- *Using ISO 56002 Innovation Management System: A Practical Guide for Implementation and Building a Culture of Innovation*; 2021

All of Harrington's books were published by McGraw-Hill, ASQ's Quality Press, Paton Press, or CRC Press. His books have been published in Chinese, Russian, Italian, Spanish, Arabic, Portuguese, French, Romanian, Hebrew, Swedish, English, and Korean.

Software by Harrington:
In 1993, he produced the interactive computer program *Benchmarking with H. James Harrington*. In 1995, in conjunction with Systemcorp, he released two CD-ROMs, *H.J. Harrington's ISO 9000 – Step by Step* and *H.J.*

Harrington's QS-9000 – Step by Step. Also in 1995, he developed a screen saver with 2000 key thoughts on performance improvement with a 50,000-word support text. In 1997, he released a CD-ROM entitled *Management Mentor.* He has also released a series of CD-ROMs related to performance improvement, as follows:

1. Creative Suspenders – 1998
2. Making Better Decisions with Numbers – 1998
3. Area Activity Analysis – 1999
4. Reliability Simplified – 1999
5. ISO 14000 and Beyond – 1999
6. War on Waste – 1999
7. Project Change Management – 2000
8. Simulation Modeling Methods – 2000

Harrington can be contacted at the following:
Email: hjh@svinet.com:
Mailing Address: 15559 Union Ave. # 187
Los Gatos California, 95032
Phone: (408) 358-2476 or (408) 356-7518

Sid Ahmed Benraouane, PhD

I like to think of Innovation Management System as the framework that helps managers discipline serendipity while creating optimum conditions for a struck of luck.

Areas of Expertise:

A leader with 20+ years of experience in multiple sectors, regions, and industries. Dr. Benraouane is a faculty at Carlson School of Management, University of Minnesota. He advises the organization on innovation, digital transformation, and AI ecosystems. He is the lead of the US ISO Working Group 1 on Innovation Management Standard ISO 56002 and a member of the US ISO/SC 42 Working Group 3 Artificial Intelligence – Trustworthiness. With a deep understanding of the economics of digital transformation in the United States and the Middle East and North Africa (MENA) region, Dr. Benraouane helps decision-makers build decision-making frameworks that enhances innovative thinking and engages the workforce.

Past Experience:
Dr. Benraouane is also a speaker at the World Government Summit on AI and Ethics (UAE 2019) and a frequent keynote speaker at regional events, such as Big Data Show (UAE), Cloud and Big Data Show (KSA), Energy Digitization Summit (UAE), Artificial Intelligence Summit (UAE), World Mobility Show and Autonomous Driving (UAE), the Middle East Military Technology Conference, and the Bahrain International Defense Exhibition & Conference (Bahrain). He is a member of the Advisory Board of the Abu Dhabi Digital Authority's Digital Next Conference.

Typical Projects:
Innovation Management Ecosystem
Innovation Management Assessment

Education:
He holds a PhD from the University of Minnesota, United States.

1

Introduction to Project Innovation Cycle

INTRODUCTION TO PROJECT INNOVATION

Julie Sweet, Accenture's CEO, manages a half-million employees that generate more than $40 billion in revenues from servicing large organizations in over 120 countries. In her role as CEO of one of the world's most respected consulting firms, she provides a unique window into what's going on in the world of business. From her unique knowledge database, the major changes in a world disrupted by a global pandemic are:

- "The first is a shift around the value of tech. We are no longer spending time asking is technology good or bad and the potential risks. Tech became the lifeline for individuals, societies, business and government."
- "The second big shift has been about speed. The most successful economies, countries, companies are those whose speed is as fast as possible. We are seeing that every day."

These statements reinforce our position that organizations around the world need to focus on generating more innovative offerings whose revenue-generating lifecycle will be shorter and shorter as new innovative offerings replace them. No organization can be content thinking that they are doing enough to keep their offerings viable. It may just be enough to squeak by today, but it is going to fall far short of meeting the needs of tomorrow. Every organization needs to look at what it is doing today to

DOI: 10.4324/b22993-1

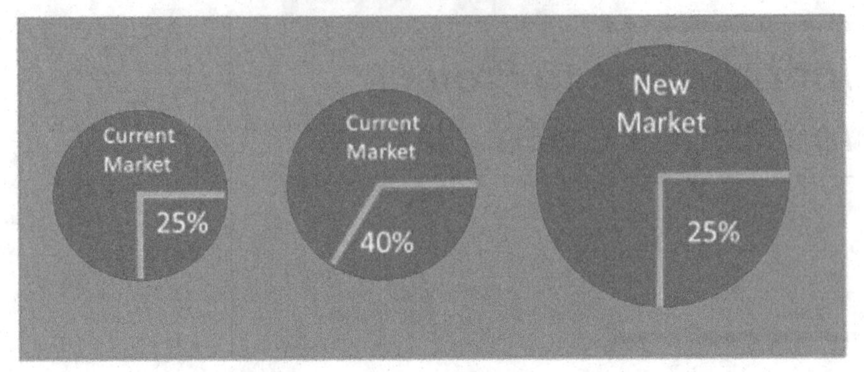
Grow Market versus
Grow Market Share
Current
Market
25%
Current
Market
40%
New
Market
25%
It can be more profitable for a larger market than increasing
the organization's share of today's market from 25% to 40%

FIGURE 1.1
Competing for Customers versus Growing the Market

keep its offerings attractive enough to bring back customers that have already been serviced as well as attract potential new customers. We can no longer survive by only trying to steal customers from our competitors. We need to focus on growing the market rather than competing for larger shares in the present market (see Figure 1.1).

Most organizations today focus on getting a bigger percentage of the market, but in the future, we are going to see a closer harmony between competing organizations working together to grow the market, rather than competing with each other to grow their market share. Growing the market provides a win–win situation for all organizations, for when you grow your market share in a stagnant market, someone else is losing his or her market share.

Basic research has become so expensive that it has forced the private industry to break away from basic research and apply their R&D resources to applied research. More and more the only ones that can afford to do basic research are universities and the government. The failure rate of new innovative product/system endeavors is estimated to run as high as 75%. I realize that from an innovation standpoint, failures are considered learning experiences. But God, do you have to make us smart that way? We need to learn more from sharing our knowledge with other people rather than depend upon self-learning by failing. There is a new saying:

- One failure – Poor management
- Two failures – Learning experience
- Three failures – Spend your time looking for a new job

What does all this mean? It means you've got to improve your PIC to significantly increase the number of new opportunities that enter your portfolio, and you have to move each innovative entity through your portfolio much faster, resulting in a higher percentage of successful projects.

TODAY'S DILEMMA

- Everyone's Talking about It

  You can't pick up a magazine, newspaper, business book, or attend a conference without a primary part of it being dedicated to the importance of improving innovation within the country and every business.
- Everyone Feels It's Necessary

  Customers want the newest, brightest, and leading-edge products and services. Companies that are the first to market continuously are the ones that are most profitable and most admired. People stand in line for hours waiting for the latest Apple products to go on sale.
- Every Organization Wants to Do It

  There's an old saying, "If you build a better mousetrap, customers will flock to your door." Certainly, all organizations realize that if they are the early leaders in their field, they will have a tremendous advantage over their competitors. The big "but" is, "How do we do it?" We are coming up with new ideas all the time but the important question is, "What do we stop doing in order to fund the resources to develop the potentially good ideas that may or may not provide value-added return?" Management wrestles with

> How much of my discretionary resources should I invest in developing new products, improve processes, artificial intelligence, basic research or in my people in order to optimize the organization's

> short- and long-range performance? My Board of Directors wants me to do more with less but that can only go so far until all the fat is off the bones and you are down to pulling off needed muscle.

But few agree on what innovation is. When is an organization innovative? I'm ISO 9000 certified by an independent organization that validates an organization as implementing continuous improvement. Does that mean we are already an innovative organization? Who do I want to feel that our organization is innovative – just my customers and my investors? If so, is it primarily product related? For productivity innovation, does it have to be something that the customer hasn't seen before or cannot get from another source? Or are you innovative when you reverse engineer your competitor's design and performance to produce your own?

ADDRESSING THE "UNADDRESSABLE"

Start out by addressing the issue that is most debated by the leaders who are documenting, teaching, advising, and implementing IMS. It certainly is very obvious that you first need to define what innovation is if you're going to improve your organization's innovation activities.

To help us get our arms around defining innovation, let us define some of the common terms that are frequently associated with the definition of innovation.

- Definition of unique: Oxford dictionary defines it as being the only one of its kind; unlike anything else. Particularly remarkable, special, or unusual.
- Definition of value added: Cambridge dictionary defines it as the total output (=value of all the products, services, etc.) in a particular region, economy, etc. after taking away the value of inputs (=materials, labor, etc.).

  To provide the reader with a more innovation-oriented definition of value added, I define it as the following: "Value added is the total combination of positive and negative impacts that a change or a new entity has on an organization and its stakeholders. It includes both tangible and intangible impacts and effects."

- ***Creativity** is idea generation*
- ***Innovation** is implementing ideas in ways that create economic value in your business world*

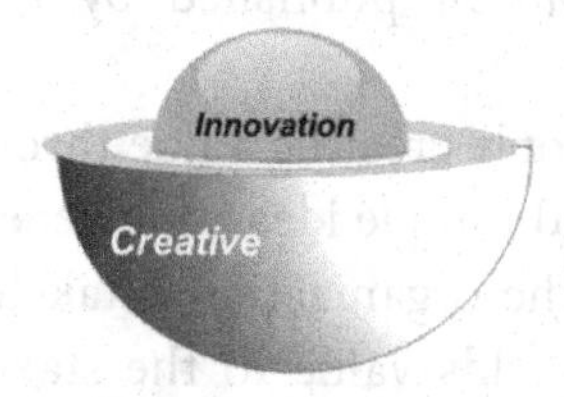

FIGURE 1.2
Magnitude of Creativity versus Magnitude of the Innovation

- Definition of creative: Being creative is using the ability of individuals to make or think of new things involving the process by which new ideas, stories, products, etc. are created.
- Definition of create: Create is the act of making something: to bring something into existence (Figure 1.2).

There are some words in the English language that are difficult, if not impossible, to get everyone to agree about their definition. Typical examples would be "pretty," "quality," and "innovation." Scholars have debated what innovation is for years. Almost without exception every book you pick up, every consultant you talk with has a different definition of innovative. Definitions run from "any time an individual does anything different, he/she is being innovative" to things such as "something that is so unique and different than anyone has done before. Innovation occurs when a new and unique idea is developed that creates positive value added to all of the organization's stakeholders."

The following are some more typical examples:

1. Innovation is a unique creative idea that is marketable.
2. Innovation is the process of creating a unique idea that is marketed.
3. Innovation is a unique creative idea that adds value to the organization's external customers.
4. Innovation is a unique creative idea that adds value to the organization's stakeholders.
5. Innovation is a unique creative idea that generates profit.
6. Innovation is a unique creative idea that when implemented provides value to the receiver of the output that is greater than the resources required to produce.

7. Innovation is the successful conversion of new concepts and knowledge into new products services or processes that deliver new customer value to the marketplace (Source: *The Executive Guide to Innovation* published by the American Society for Quality).
8. Innovation is people creating value through the implementation of new, creative, and unique ideas that generate combined measurable added value to the organization's stakeholders. Innovation is how an organization adds value to the stakeholders by implementing creative new ideas.
9. Many dictionaries commonly define innovation as the "carrying out of new combinations" that include "the introduction of new goods, … new methods of production, … the opening of new markets, … the conquest of new sources of supply … and the carrying out of a new organization of any industry" (Source: Wikipedia).
10. According to ISO Technical Committee WG 2 that wrote 56000:2020 Innovation Management Systems – Fundamentals and Vocabulary Standard: Clause 4.1.1 Innovation is a new or changed entity realizing or redistributing value. (©ISO. This material is excerpted from ISO 56000:2020, with permission of the American National Standards Institute (ANSI) on behalf of the International Organization for Standardization. All rights reserved.)

    There's a saying, "A camel is a horse designed by a committee." This is probably a little true when applied to ISO 56000 Standard's definition of innovation. But its definition is the one that we should be using as it is the agreed-to standard in most major countries around the world from Russia, to China, to Japan, the United States, to Germany. It's particularly good because it's broad enough to cover most of the definitions although it may include some conditions that specific individual definitions would not classify as being innovative.
11. After many long conversations with the ISO Technical Committee 279 – Innovation Management Systems, studying and analyzing the previously presented definitions of innovation, and research related to a number of papers and conferences we attended, we came up with the following personal definition of innovation:

Innovation is a new or unique idea or concept that adds value to the organization and its stakeholders. Innovation is the act of taking a new, unique, and creative idea, developing and funding it, producing it, and distributing it to external customers that results in creating value to the organization, the consumers, and the organization stakeholders.

> *You can have creativity without innovation. You cannot have innovation without creativity.*
>
> **—H. James Harrington**

FACTORS AFFECTING INNOVATION

There are many factors that need to be considered when you apply the word "innovation" to the activities that go on within an active worldwide environment. Some of them are:

- Does it apply to only the perception of the end-user/general public?
- Does it include continuous improvement that is standard on most entities?
- Do we have to consider negative, as well as positive, impacts on all the organization stakeholders?
- How does it apply to the entertainment industry (sports, music, art, movies, TV, etc.)?
- How does it apply to not-for-profit making organizations?
- How does it apply to the military?
- How does it apply to government offices?
- How does it apply to education?
- How frequently do you have to do something innovative to be considered an innovative organization?
- Is every change within the organization an innovative change? If not, how do we know which ones are not?
- What parts of the business should be focused on to improve the organization's innovation?
- How do we translate intangible gains like reduced cycle time, lives saved, increased customer satisfaction, improved morale, impact of

culture change, etc. into a dollar value that can be considered when you're calculating return on investment?

THE FIVE TYPES OF INNOVATION

There are five types of innovation based upon the output from the project/program:

1. Product Innovation, which primarily leads to upgrading and producing a current product so that it has a competitive advantage over other products that will be available. In other cases, it results in the delivery of a new product that is not available at the present time.
2. Process Innovation, which primarily leads to producing a competitive advantage through lower prices or reduced cycle time or improved reliability or a combination of all three.
3. Sales and Marketing Innovation, which primarily leads to producing a competitive advantage or a marketing mix (target market, distribution, product, price, and promotion).
4. Management Innovation, which primarily leads to producing a competitive advantage through better organizational ways of achieving the organization's goal or better use of the organization's resources.
5. Service Innovation, which applies to servicing the customer before and after they have purchased the items and services to internal organizations. This basically drives improvements in responsiveness, understanding of the consumer's environment, and relationships.

The Three Subcategories of Innovation

For each of the five types of innovation, there are three subcategories:

- Breakthrough Innovation

  This is when a radical new design or approach is created and implemented. It's often referred to as *A-ha* innovation.

- Evolutionary Innovation

  This is when a logical expansion of an existing product, process, or strategy is implemented. It typically does things like making it smaller, making it faster, and adding additional capabilities.
- Gradual Innovation

  This is when many small improvement activities are implemented throughout the organization. This is commonly called "continuous improvement" in the product and employee's performance. It can be as simple as a janitor finding a new broom that sweeps better, to the project engineer that uses the next-generation technology to perform the same or similar function at the present one.

The five types of innovation are sometimes called "Five Levels of Innovation." Many of our experts argue continuously about whether an activity is a continuous improvement or innovation. Others argue that continuous improvement is part of innovation. In support of the separation of continuous improvement and innovation, ISO 56002:2019 Standard for Innovation Management Systems only specifies that continuous improvement can be applied to the IMS itself. But if you look at it in the broadest sense, any change that adds value to the organization or the customer is a form of improvement.

The improvement opportunities can be broken down into five levels. Based upon a massive study of thousands of patents and technology systems, Genrich Altshuller categorized improvement opportunities as given in Table 1.1.

Figure 1.3 displays the five types of improvement in a graphic format. We found this knowledge on the shocking side, but it was very interesting. It indicates that 95% of all ideas that are patentable fall

TABLE 1.1

The Five Types of Improvement

Type 1 – Apparent Solutions (Gradual Innovation)	= 68.3% of the changes.
Type 2 – Minor Improvements (Gradual Innovation)	= 27.1% of the changes.
Type 3 – Major Improvements (Evolutionary Innovation)	= 4.3% of the changes.
Type 4 – New Paradigm (Breakthrough Innovation)	= 0.24% of the changes.
Type 5 – Discoveries (Breakthrough Innovation)	= 0.06%.

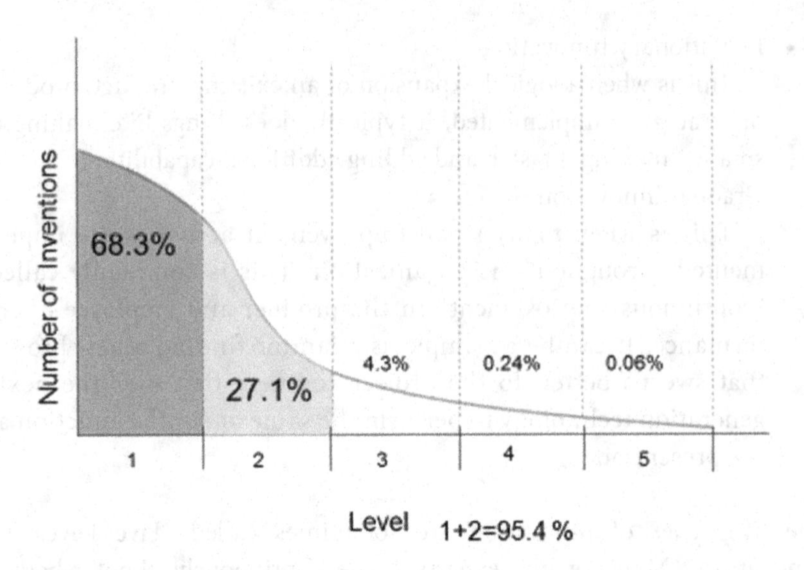

FIGURE 1.3
Five Types/Levels of Improvement

into the evolutionary or gradual changes to our products, processes, and services.

- Type 1 – Apparent Solutions:
  Is Apparent Solutions more continuous improvement, or is it new and unique so it can be classified as innovative? We would estimate that the Three Sigma of this population would be classified as continuous improvement, not as innovation.
- Type 2 – Minor Improvements:
  What percentage of the Minor Improvements is continuous improvement, and what percentage is innovative? Based upon looking at a small sample of data, we would estimate that 80% to 85% of the minor improvements are continuous improvement and only 15% could be classified as innovative.
- Type 3 – Major Improvements:
  What percentage of the Major Improvements is continuous improvement, and what percent is innovative? We estimate as little as 20% to 30 % are continuous improvement and as much as 70% to 80% are innovative.
- Type 4 – New Paradigm:
  What percentage of the new paradigms is continuous improvement, and what percentage is innovative? Here a whopping 95% plus

of the patents would be considered innovative and less than 5% of the opportunities would be considered a continuous improvement.

- Type 5 – Discoveries:

  What percentage of the discoveries is continuous improvement, and what percentage is innovation? Again, this ratio is more than 99% for innovative and less than 1% for continuous improvement.

CONCLUSIONS RELATED TO TYPES OF INNOVATIONS

You may not agree with the exact ratio that we are estimating, but it should give you an idea of the magnitude of the continuous improvement initiative's part of the normal day-to-day continuous improvement activities and innovation for each of the five categories. More important is to point out that approximately 95% of all the improvements that are new and creative fall in the categories of apparent solutions and minor improvements. This would indicate that the innovative system needs to focus upon 5% of the total improvements that are made up of Types 3, 4, and 5.

It's also important to point out that these are the types of changes that give an organization a very significant competitive position and reputation. Often a Type 4 change will be the trigger that promotes many of Type 1 and 2 changes. Type 1 and 2 changes require a much less sophisticated IMS that is less expensive to operate. Most organizations exist by focusing all their improvement activities on Types 1 and 2 of changes, allowing competitive organizations to focus their attentions on Types 3, 4, and 5 of opportunities. This greatly reduces their R&D costs and their project failure rates. The disadvantages put them in a follower-type mode rather than a leadership role. Frequently they try to make up for this by transferring research and development funding into the sales and marketing campaign.

COMMON CREATIVITY INNOVATION KILLERS

(Source: Book entitled *Creativity Toolkit*, H. James Harrington)

Are you an innovation killer? Do you encourage or repress creative and innovative thoughts? A newly hatched idea is fragile. We need to encourage it, cultivate it, and help it to develop and grow, not casually discard it. The following are 12 commonly-used phrases that discourage innovative thinking:

- It won't work.
- It makes me afraid.
- We tried it already.
- That can't be done.
- It will never work here.
- Let's be serious.
- That's ridiculous.
- What's original about that?
- How dumb can you be?
- You obviously don't understand the situation.
- That's a silly idea.
- That's impossible.

We can't build a better tomorrow by using yesterday's methods. Businesses that expect to make it in today's global marketplace must begin by tapping the innovation of all employees, not just a few maverick inventors or dynamic CEOs. Competitive advantage today comes from continuous, incremental innovation.

—Harold R. McAlindon

Now here are 13 quick techniques you can use to overcome the above-mentioned put-downs and turn on your innovative powers:

- Create innovative mental pictures in your mind, and turn these pictures into reality.
- Keep your mind open to new ideas by presenting new experiences to your senses. Be a keen observer of the environment that you come in contact with. Provide your mind with the raw materials that it needs to be creative and innovative.
- Do something creative each day. Set aside a specific time each day to review the creative and innovative things that you accomplished.
- Gather data to prove you were right. Focus your creativity and innovation on simplifying the old and new approaches.

- Maintain a questioning attitude. Remember, there is always a better way and if you don't find it, someone else will and use it as their stepping stone to get ahead of you.
- Don't be afraid to take a risk. You will never fulfill your true potential if you play it safe.
- Record your ideas as soon as you get them. Keep a notepad with you at all times.
- Take time to relax and unwind. Take a long walk or a long hot bath. Play golf or restful music. Try meditation or yoga.
- Don't accept limiting factors as being unchangeable or correct.
- Gain confidence and enthusiasm by first focusing your innovative effort and ideas on things that are within your control to implement.
- Help others to be innovative by pointing out the good points related to their ideas, not the bad points. We already have too many devil's advocates. Be an angel's advocate.
- Find your creative/innovative time of the day. Some people are morning people. Others are evening people. We all function differently. Sample your emotions and creative/innovative powers to determine when you are the most creative and innovative. Then, set that sacred time aside to work on developing new concepts.
- Start today to improve your creative and innovative processes. It has been said, "Yesterday is history, tomorrow is a mystery. Today is a gift. That's why it's called the present."

I am a great believer in luck, and I find the harder I work, the more of it I have.

—Stephen Leacock

YOUR CREATIVE AND INNOVATIVE POWERS

As we begin to review the many techniques you can use to turn on your creative and innovative powers, here are some affirmations about you and others like you that serve as a foundation for our ideas:

- We are confident that you are or can be creative and innovative.
- We are confident that you can improve your creativity and innovation. It has been estimated that Leonardo DaVinci and

Thomas Edison used less than 50% of their potential creativity capabilities.
- We are confident that the regular use of the "mind expanders" defined in this book will improve your innovation.
- We are confident that risks, innovation, and rewards go hand-in-hand.
- We are confident that creativity and innovation will become even more critical to real success in the 21st century than ever before.
- We are confident that creative people get more joy from life.
- We are confident that if you do not use your creative and innovative powers, you will become less capable of using them.
- We are confident that real success goes to creative and innovative people who can implement their ideas and concepts.

The things that get done are the things that are easy to do and where the rewards are the highest.

—H. James Harrington

The 10 Ss

Successful, growing companies have found that the answer to many of their problems is to become more creative and innovative. Of course, that's easier to say than to do. As a result, we have focused on defining the innovation drivers. We start out by focusing on the tried and proven McKinsey 7S model:

1. Shared vision
2. Strategy
3. Systems
4. Structure
5. Skills
6. Styles
7. Staffing

Now Dr. Harrington will admit that he is not a fan of "McKinsey," but after working over 10 years with Ernst & Young, he got to really know the McKinsey organization, which was his competitor. He admits that McKinsey had some brilliant creative people working for it. When we started comparing these 7 Ss, which we will refer to as performance drivers, to the activities going on in some of the best-known innovative organizations, we found that all seven were relevant and important performance drivers that must be considered and addressed in establishing an innovative organization.

As we gained more experience in innovation transformation, we found that there were three other performance drivers that needed attention to keep pace with the fast-changing technology environment/competition organizations face today. They are:

- Specialized technology – information technology systems
- Systematic change management
- Strategic knowledge management

Adding these 3 Ss to McKinsey 7 Ss makes up a new grouping called 10 Ss. These are subdivided into hard drivers and soft drivers. Hard drivers are easy to define and the organization can directly influence them. The hard drivers are:

1. Strategy
2. Structure
3. Systematic change management
4. Systems
5. Specialized technology

The soft drivers are less tangible and more influenced by culture. Both the soft and hard drivers have a major impact on an organization's performance. The soft drivers are:

6. Shared vision
7. Styles
8. Staff
9. Strategic knowledge management
10. Skills

Clauses 1, 2, and 3

The content in Clauses 1, 2, and 3 is obvious and does not need further description about what is contained in these clauses.

Knowledge Management System

As you will notice, the 10 Ss foundation is a very holistic view of the total organization and the habit patterns with management and employees. To accommodate this and to take advantage of the synergy between each of the 10 Ss, we view their output as flowing into an information hub. This combination of the 10 organizational drivers and the information hub make up a key part of the organization's Knowledge Management System (KMS) (see Figure 1.4).

As we have previously stated, all organizations already have an IMS of some kind in place, but today many individuals or organizations are considering or in the process of upgrading it. Obviously, this book is designed to help the individual projects/programs/processes through an already-established IMS.

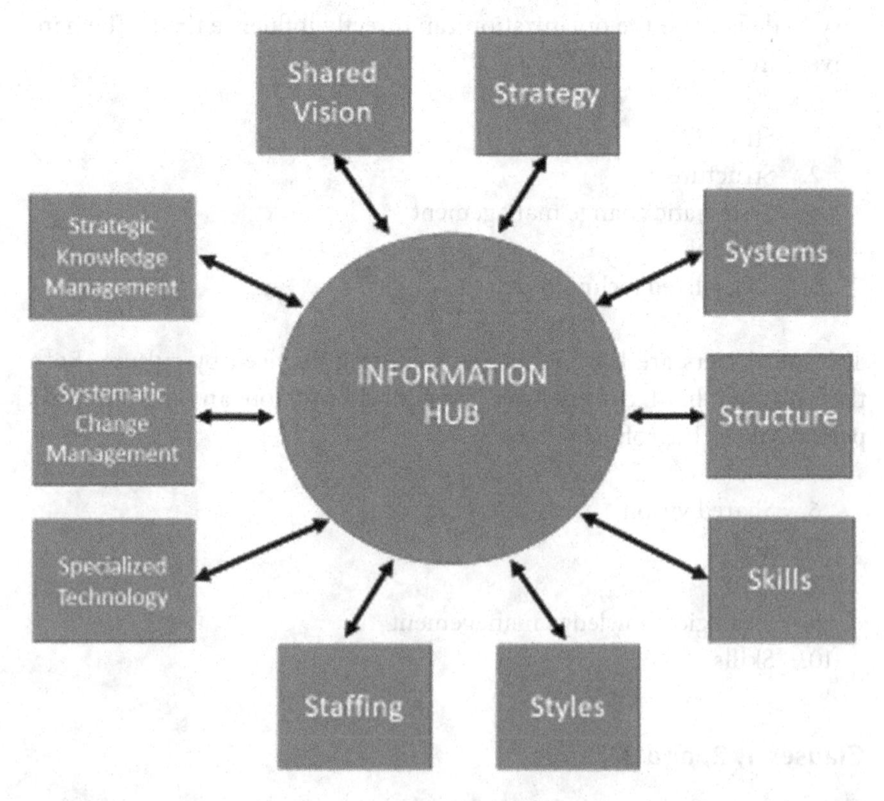

FIGURE 1.4
Key Parts of the Knowledge Management System

Unfortunately, the bouquets of flowers are given to the individuals that designed and implemented a new or upgraded system and not to the legion of employees that make the system perform in spite of many problems and bottlenecks. If we were to liken it to a horse race, the system design and implementation was like providing a racetrack with all its hurdles and challenges left in place. The winners are not the individuals that designed and installed the racetrack, but it is the jockeys and horses that use the racetrack and compete to win the race. Likewise, you do not win the race by providing the system, but you win the race based upon how the system is used.

If you are reading this book, you probably are an individual who has been assigned to manage an important project that will have a significant impact on today's survival and tomorrow's bright future. Now the value of what the system designers created is put to the ultimate evaluation of designers' and implementers' skill and imagination or the lack thereof. Today's good is not good enough; only our best will stand up and meet the challenges brought about by our international competition.

We have been fortunate to work with a number of organizations that implemented Six Sigma improvement approaches. In each case, the basic systems were almost identical, and certainly the foundations they were built upon were identical. But the results varied from outstanding to so-so. For a while I had a hard time understanding why they didn't get as consistent results every time they put in identical systems. I eventually realized the key to real performance is not the system but how innovatively the system is utilized.

- An outstanding system poorly utilized gives you unsatisfactory results.
- A minimum system creatively utilized produces good results.
- An outstanding system creatively utilized produces outstanding results.

2

ISO 56002:2019

INTRODUCTION

This book was designed to improve your organization's innovative projects and programs' value added by reducing processing time, risk of failure, and costs while improving performance and customer satisfaction. It was designed to work with any level of IMS that your organization presently is using. It is designed to work with the basic system to decrease the risk of project and program failures, reduce the cycle time from opportunity to delivery, and increase the number of opportunities available to the organization. If you have an effective IMS in place already, you will find this book provides valuable content to help you focus on reducing variation and continuously improving the value-added content of your programs and projects. It offers a systematic approach that should provide the controls required to mitigate risks related to schedule slippage, cost overruns, and unapproved added scope. This basically is the management controls over the creation, development, implementation, and operational processes.

If your IMS is performing so poorly that you want to completely restructure your organization to make it much more innovative, you really need to work with the ISO International Standards 56000:2020 set of documents. At a minimum, you'll need to generate a basic cultural change that usually is difficult to implement. So why not get the advice from the innovation leaders around the world at a cost that is less than you'd pay for a one-hour meeting with any one of these individuals? We are fortunate that the International Organization for Standardization took the time and energy to bring together the thought leaders in innovation to share their innovation knowledge with you in the form of the ISO 56000:2020 set

DOI: 10.4324/b22993-2

of Innovation Management System Standards. Consider the suggestions made in two of H. James Harrington's books – *Using ISO 56002:2019 Innovation Management Systems: A Practical Guide for Implementation and Building a Culture of Innovation* (published by CRC Press 2021) and *TIME – Total Innovation Management Excellence* (published by CRC Press in 2020). With the aid of these books, your organization can develop an effective system that can repeatedly produce innovative entities in the quantities required to maximize the value added to the organization, its investors, and its customers.

If you have an IMS that needs to be improved, this book, together with ISO 56004:2019 Innovation Management Assessment, provides a methodology that will minimize the time and effort required to make the necessary changes. In fact, if your IMS is obsolete and you need to establish a new culture or if your present IMS is good but you would like it to be better, then applying the techniques that are defined in this book to the individual entity is a basic necessity.

TIME

The TIME methodology is designed to develop the culture that will allow innovation to prosper and grow within an organization, and this book is designed to maximize the value-added potential of the culture. It's like TIME technology combined with ISO 56002:2019 is the car, and proper use of the PIC is the gasoline that gets you from place to place. Surely you could push the car by hand or harness it to a horse to move the car, but that is very time-consuming and undesirable, and by the time you got home, the ice cream would have melted. The good thing about the PIC is that its approach to applying proven advanced innovation methodologies to individual projects/programs is the best and fastest way to create your desired result.

This book is not intended to provide the reader with the information and skills required to effectively implement ISO 56002:2019 and create an innovative culture across the total organization. However, it does provide you with the methodology and techniques that will bring about significant improvement in the individual projects and programs that it is applied to. Simply put, we suggest that you need to consider the suggestions provided

TABLE 2.1

ISO 56000 Standards

ISO 56000:2020	Fundamentals and Vocabulary
ISO 56002:2019	Innovation Management System
ISO 56003	Tools and Methods on Innovation Partnerships
ISO 56004:2019	Innovation Management Assessment
ISO 56005	Tools and Methods on Intellectual Property Management
ISO 56006	Tools and Methods on Strategic Intelligence Management
ISO 56007	Tools and Methods on Idea Management
ISO 56008	Tools and Methods on Innovation Management

in this book, and those that are presented in the ISO 56002:2019 Standard along with the instructions in the TIME methodology, in order to bring about a major innovation cultural change. If you want to improve the IMS, we suggest you need to use the approaches defined in ISO 56004:2019 and the information and processes described in the book *Using ISO 56002:2019 Innovation Management Systems: A Practical Guide for Implementation and Building a Culture of Innovation* (published 2021 by CRC Press).

ISO's Technical Committee 279 is one of the most active committees in the program and, as a result, has accomplished a great deal in the past two years. At the present time, they have five active working groups and at least one other being considered. Table 2.1 provides a view of the working groups and their assigned tasks.

ISO 56000:2020 STANDARDS

The focus of this book is directed at helping the individual project, program, or method gain maximum value from the IMS the organization is using and to assist in the implementation and transformation of the major desired changes. Because of this, it is not practical for us to try to acquaint you with the intricacies and risks related to upgrading your culture to be more innovative through the use of the ISO 56000:2020 Standards set and the use of the TIME methodology. But it may be helpful to provide you with some introductory perspectives if you have a basic understanding of what is contained in the ISO 56002:2019 Innovation Management Systems Standard.

ISO 56002:2019 STANDARD

We have found that one of the most difficult parts of the PIC is getting it started. We want everyone from the chairman of the board to the floor sweeper to be continuously searching for improvements or new opportunities that are worthwhile implementing. Once a potential improvement opportunity has been defined, we need to get ready to answer the often-unspoken questions customized to every individual's perspective: Questions like, What is it going to do for me? How do I identify improvement opportunities? What kind of opportunities is management looking for? Why are we doing it? Why do I have more work than I can handle now? What do you want me to not do that I'm doing now? How will it affect the people I work with? If I identify an improvement opportunity, will it be addressed, etc.? In these cases, the standard organizational answer to the question is a poor answer. We have to answer each question with an answer that reflects the individual's job, emotions, worries, hopes, and fears. Certainly, the management team needs to prepare themselves so that they can provide employees with encouraging answers to their emotions.

To answer these and similar questions related to identifying an improvement opportunity, let's consider some related factors.

1. Have we made innovation and creativity part of every employee's job responsibility? This can easily result in hundreds of improvement opportunities being identified each month. Have we implemented a system to handle them?
2. Have we established a team performance culture that is used throughout the organization?
3. Has everyone within the organization gone through basic problem solving training?
4. Have we screened everyone on how to do a simplified added value-type analysis? Key personnel have had detailed training on how to calculate the value added for high-leverage innovative opportunities.
5. Have we established an innovation project review team?
6. Why is our project failure rate running at about 15% failures?

We also need to consider the type of innovation opportunity that we want to be identified. Think carefully about your system. Is it capable of handling an innovative culture? If not, you are making it difficult for your project managers to be successful. If your innovation management system needs to be updated, you need to consider some of the advice that was given by an international team for the International Organization for Standardization (ISO). Their standard 56002:2019 Innovation Management System – Guidance provides many suggestions that you will find helpful and valuable (see a short summary of the standard in the next chapter).

However, the first thing you need to understand is that there is a big difference between the meanings of the words "should" and "shall" in the ISO dictionary.

Following definitions outline the difference between "should" and "shall" in ISO Standards:

1. Definition of should–Should is used to indicate a goal which must be addressed by the design team but is not formally verified. Goals, non-mandatory provisions.
2. Definition of shall–Shall is used to indicate a requirement that is contractually binding, meaning it must be implemented, and its implementation verified. Period! Don't think of "shall" as a word, but rather as an icon that SCREAMS: "This is a requirement." If a statement does not contain the word "shall," it is not a requirement.

ISO 56002:2019 is a "should do" type book that places no requirement for any individual to implement clauses in this standard. The choices of the clauses and how they are implemented is left to the discretion of the organization that is using the IMS to manage their organization. We particularly like this approach as it leaves the decision-making to the individuals that best understand the organization's mission, goals, priorities, commitments, culture, and commitments.

ISO 56002:2019 Organizational Structure is made up of 10 clauses. They are:

1. Scope
2. Normative references
3. Terms and definitions
4. Context of organization
5. Leadership
6. Planning

7. Support
8. Operation
9. Performance evaluation
10. Improvement (©ISO. This material is excerpted from ISO 56002:2019, with permission of the American National Standards Institute (ANSI) on behalf of the International Organization for Standardization. All rights reserved.)

Clause 4.0 – Context of the Organization

You should consider all internal and external stakeholders when developing your IMS. In Clause 4, we discuss the following four major considerations:

1. Defining the organization's status and key issues
2. Identifying stakeholders' needs and expectations
3. Defining the requirements and limitations of the IMS (scope)
4. Defining the content of the IMS

Overview:

- Defining the extent that internal and external issues impact the IMS
- Understanding the interest and needs of the organization's stakeholders (interested parties)
- Presenting cultural best practices
- Considering the previous three conditions to develop the scope of the IMS

Purpose:

In IMS, they use the word "context" extensively in this Clause.

- Definition of context: "ISO 56000:2020 defines context as the combination of internal and external issues that can have an effect on the organization's approach to developing and achieving its objectives. It is also sometimes called organizational environment." (©ISO. This material is excerpted from ISO 56000:2020, with permission of the American National Standards Institute (ANSI) on behalf of the International Organization for Standardization. All rights reserved.)

- Definition of leadership: Leadership is a process of social influence, which maximizes the efforts of others, toward the achievement of a goal.

  Note: Notice key elements of this definition. Leadership stems from social influence, not authority or power. Leadership requires others, and that implies they don't need to be "direct reports." Leadership is not defined in ISO 56000:2020.

Most of the items discussed in Clause 4 are basic items that apply to most of the systems that are embedded in most organizations. The exception to this is technology advancements that are applied only to a few parts of the organization even when they only apply to the continuous evolution-type improvements within an already-established product. For example, the cultural aspects that support the PIC are the same that support the personnel systems.

If you consider the PIC starting with marketing, development engineering, product engineering, manufacturing engineering, quality engineering, manufacturing, procurement, and sales, it would be very undesirable and costly to have a separate culture for the innovative activities within the organization that differ from the rest of the operation's culture. The only exception to this would be when management establishes an organization that is allowed to take on a different set of principles and operating procedures. For example, you may build typewriters in one plant and manufacture computer chips in another location. Or there may be a very different culture in research and development than there is in finance. We suggest that Subclauses 4.1, 4.3, and 4.4 be viewed from a total organization's viewpoint, not just the innovation cycle.

Clause 5.0 – Leadership

Purpose:

The purpose of this clause is to provide a general framework to the role of leadership to enhance the IMS. It discusses the components of leadership behavior that inspire people to engage in innovation activities and programs and describes actions leadership should take to support the IMS. Examples of actions are the communication and awareness campaign that leaders should engage in, executive presence, and the structure that top management should provide to create an alignment and coherence in the functioning of the IMS.

Clause 5 consists of the following subclauses:

- Leadership and commitment
- Innovation policy
- Organizational roles, responsibilities, and authorities ()

Overview:

This chapter focuses on the leadership behavior that supports an IMS. The chapter shows you how to create a vision, how to craft an innovation strategy, how to communicate it to employees, and how to set an innovation policy that clarifies roles and responsibilities.

Clause 6.0 – Planning

Purpose:

The purpose of this clause is to provide guidance for organizations that are setting up an IMS. This clause is directed at accomplishing two objectives. First, it defines what is required to plan for installing or upgrading an IMS, and second, it defines how individual projects/programs will be processed through the IMS. The major purpose of this clause is to determine if an initiative will become part of the organization's portfolio and thus have resources committed to it or to drop the initiative.

A final reminder: The ISO Standard 56002:2019 and this book were written to provide guidance for midsize and large organizations. Applying some of the standard and the tools, methodologies, and some of the culture-building recommended in this book to a small startup company could easily burn up (use up) resources which could be used much more effectively doing other activities. After all, it is not the IMS that is important; it is how it can be used to be of more value to you, your organization, and the other stakeholders.

Overview:

This clause focuses on the process of developing a plan that will be used for assigning resources (for example: employees, money, space, and management resources). It addresses establishing a plan to develop, install, and evaluate an IMS. It also provides guidance on planning for the activities that could be used in processing an innovative improvement through the PIC.

It consists of four subclauses. They are:

- Subclause 6.1 Actions to address opportunities and risks
- Subclause 6.2 Innovation objectives and planning to achieve them
- Subclause 6.3 Organizational structures
- Subclause 6.4 Innovation portfolios ()

Clause 7.0 – Support

Purpose:

This clause provides guidance on how to put together the resources needed to establish, implement, maintain, and improve a support system for your IMS. Here you will learn how to use resources and capabilities, understand the importance of managing and engaging people and teams, and how to secure funding to your innovation activities and programs.

Overview:

This chapter discusses some of best practices on how to establish a support system for innovation programs and activities. It discusses the role of people, rewards, knowledge, and infrastructure, such as innovation labs, in enhancing the performance in an IMS. It also describes how organizations should address collaboration and change management, as well as IP issues. Finally, this clause discusses some known tools and methods that are used in the industry to engage teams in creativity and innovative thinking. It consists of the following subjects:

- Resources (people, time, knowledge, finance, infrastructure)
- Competence

- Awareness
- Communication
- Documented information
- Tools and methods
- Strategic intelligence management
- Intellectual property management

Clause 8.0 – Operations

Purpose:

To provide guidance on how to manage the innovation process and how to orchestrate different operations that are needed to support the innovation process. This clause describes the control mechanisms of collaboration, criteria by which you define innovation, and the agile governance framework that you need to be put in place to ensure a more fluid innovation process.

Overview:

This clause focuses on how to manage innovation initiatives and how to prepare the innovation process, i.e., create a concept, identify opportunities, validate concept, and develop and deploy solutions. It also provides tools and best practices on how to evaluate an idea at the end of each step, and provides a risk assessment matrix at the end of each step. It also helps you understand what activities are needed to operate your PIC while identifying some best practices on using the ideation process. This clause consists of the following subjects:

- Operational planning and control
- Innovation initiatives
- Innovation processes

Clause 9.0 – Performance Evaluation

Purpose:

This clause describes the process by which you evaluate the performance of the IMS. It provides the general framework to help you create and implement a monitoring system that allows you to measure, analyze, and evaluate the performance of innovation activities and programs.

It describes the steps needed to establish the monitoring system and discusses tools and methods needed to evaluate the IMS. It also provides a discussion of the difference between qualitative and quantitative methods to help you decide about the types of metrics you need to create to install the evaluation system.

Overview:

This clause on the general framework allows you to put together a coherent evaluation system, know how to link it to your internal audit, and know an annual management review process. It also shows you how to align these three actions to create a coherence in the functioning of the IMS evaluation.

- Monitoring, measurement, analysis, and evaluation of the IMS
- Internal audit
- Management review

Clause 10.0 – Improvement

Purpose:

The purpose of this clause is to provide guidance on how to implement an improvement system and manage the different steps that link it to the shortcomings and the weaknesses identified by the IMS evaluation process. It helps you understand how you can install corrective actions to stop immediately the deviation, and establish a preventive action plan to avoid future problems.

Overview:

The focus of this clause is on the steps you need to take to install a leading system that helps you identify the weaknesses and how to prevent them from happening in the future. At the end of the clause you will learn how to conduct an improvement strategy based on output from your evaluation system. You will also learn how to communicate with different parties and stakeholders internally to improve the IMS classification.

There are two components to this clause:

- Deviation, nonconformity, and corrective actions
- Continual improvement

MORE DETAILED VIEW OF SECTION 8: OPERATIONAL

We have expanded Clauses 8.1, 8.2, and 8.3 to provide the reader with a better understanding of the meaning of these clauses and a better view of the index thinking that went into preparing ISO 56002:2019. These three clauses were randomly selected because I was working on this expansion for one of my clients and I was preparing the Level III write up.

Section 8.0 is a *should-type* section. It is made up of the following clauses:

- Clause 8.1 Operational planning and control
- Clause 8.2 Innovation initiatives
  - Clause 8.2.1 Innovative initiatives should be managed considering 12 different items
  - Clause 8.2.2 Innovative initiatives should be implemented using a single or combination of six different approaches
- Clause 8.3 Innovation processes

Clause 8.1: Operational Planning and Control

The organization should plan, implement, and control innovation initiatives, processes, structures and support needed to address innovation opportunities, meet requirements, and to implement the actions determined in 6.2. (©ISO. This material is excerpted from ISO 56002:2019, with permission of the American National Standards Institute (ANSI) on behalf of the International Organization for Standardization. All rights reserved.)

Operational control is about the way you apply and administer different systems, processes, and metrics to manage efficiently the innovation system and align better different organizational resources to create the harmony and the coherence needed for the innovation ecosystem to function in an optimum manner. To do that, Subclause 8.1 of ISO 56002:2019 suggests that you pay attention to external changes happening around the organization (Clause 4, the Context of the Organization) and different planning processes discussed in Clause 6 of the standard (Planning) to be able to integrate change, address risk, and think of opportunities when you manage your operations.

Furthermore, and to be able to manage well IMS operations, Subclause 8.1 suggests that you pay attention to six actions:

- The first is criteria by which you define innovation. Defining innovation is a critical step to aligning people and behavior around a common philosophical interpretation of what innovation is. It helps you also adopt a similar language that clarifies the value, the goals, and the intent. Clarity of conceptual issues, especially for a relatively new topic such as innovation management system, affects the way people perceive things and creates a better efficient thinking process. It also creates confusion about what the company means by innovation. As we argued in Chapter 1, we will be using the innovation definition defined in ISO 56000:2020 – Innovation Management – Fundamentals and Vocabulary because it is the one agreed to by the International Organization for Standardization. The big advantage to this is to provide such a large scope definition that covers most of the more refined definitions. It obviously was the best compromise that the International Committee could agree on.
- The second action recommended by Subclause 8.1 is to define the control mechanics by which you manage different operations.
- The third action recommended by Subclause 8.1 to manage innovation operations is the documentation. Of course, documentation is a requirement of ISO 9001:2015, and it needs to apply to your IMS as well. Documenting different operations, as well as updating the documentation when things change, helps you track the evolution of the system, the learning, and the improvements of your innovation operations.
- The fourth action Subclause 8.1 recommends is change management. ISO 56002:2019 suggests that you define a framework that allows you to take advantage of changes in the external environment while mitigating its adverse impact on your operation.
- The last two actions ISO 56002:2019 recommends to manage operations are collaboration and agility. Collaboration and agility are critical in managing innovation operations, and they are relatively new concepts. We will discuss them in more detail.

Collaboration and Partnership

In innovation, collaboration and partnership are a must. There is no way for a company today to innovate without being engaged in a network of collaboration where participants share information, data, and research. So, when you engage in a collaborative arrangement, it is critical that you take control of that arrangement and get involved in different operations. ISO 56002:2019 suggests that when you outsource your collaboration initiatives, you need to make sure that you are in charge of managing the agreement and that you are fully involved in that partnership by creating control tools and gating mechanisms that help you to manage different operations.

A good framework to use to implement collaboration and partnerships is the ISO 56003:2019 Innovation Management – Tools and Methods for Innovation Partnership – Guidance. This standard is part of the ISO innovation series and can be helpful in creating and managing collaboration and partnerships.

Generally speaking, to implement an innovation collaborative initiative, you need to take three important steps: Decide whether to enter an innovation collaborative initiative; identify and select partners; and align the perception of the value. Let us look at these steps one by one.

- Decide Whether to Enter into an Innovation Collaboration Initiative

  There are four reasons for a company seeking to develop its innovation capabilities to enter into a collaborative agreement. The first one is complementarity. The reason you would want to enter into a collaborative agreement is to get easy and free access to resources and capabilities that may add value to your current resources and capabilities. Complementarity is an important reason for collaboration. You may have a strong borrowing capacity, but you lack customer insights. You may have an important stock of technology, but you do not have a scientific research capability. And finally, you may have an attractive physical location, but you do not have access to talents. These are situations in which complementarity may help overcome resource constraints in developing an innovation ecosystem.

  So, before you decide to enter into a collaborative agreement, ask yourself the following gating questions:

  - Gating question 1: Will the partnership add value to my innovation capabilities and in what ways? If the alliance does not add any value to my innovation ecosystem, then the partnership does not make sense.

  - Gating question 2: What type of synergy will I develop by creating the partnership, and how will the synergy boost my innovation capabilities?
  - Gating question 3: Will there be a cultural fit between the two organizations that will boost collaboration, creativity, and cohesion between the teams?

- Select and Identify the Right Partner

  The second step in managing collaboration is to identify and select the right partner. ISO 56003:2019 provides a long list of potential organizations that can be a good target, but generally speaking, partners can be identified from four different main categories: R&D laboratories, business clusters, competitors, and customers. In this step, you need also to ask three gating questions:
  - Gating question 1: Is there enough trust between the two organizations so teams can work openly and share information and knowledge in a way that boosts my innovation ecosystem internally?
  - Gating question 2: Have we addressed the issues that relate to intellectual property rights in order to manage innovation input?
  - Gating question 3: Is there a cultural alignment between the two organizations that would allow us to align our perception of value in the collaboration?

- Manage the Partnership Interaction

A successful partnership is one that relies on good management practices. You need to put in place mechanisms that allow the partnership to run smoothly and solve practical issues that may arise. You need also to establish control systems, whether these control systems are financial, strategic, or structural in order to be able to reap the benefit of the partnership.

Finally, and as we have been arguing throughout this chapter, innovation is an inherent human activity. It relies on trust and openness to flourish. If trust is not there, the collaborative agreement runs into issues that manifest themselves later in the form of cultural clashes. If you have not conducted cultural due diligence in order to test the culture, the perception, and the values of your partner, it may be difficult to create a smooth partnership that helps you boost your innovation operations. As well said by Alastair McLeod, CEO of Veracity:

Culture defines how people work together to achieve business goals. This has particular relevance when it comes to innovation: It can be hard to launch a new product or arrive at a new scientific breakthrough against a backdrop of misunderstanding or even mistrust. A poor cultural fit can be highly disruptive, throwing projects off the track and delaying crucial decisions.

- Organizational Agility

  Agility, as a conceptual framework, refers to the ability of the organization to respond quickly to market changes, while still focusing on your innovation projects and operations. The concept of agility has been defined by different authors. Aaron Smith (2015), a principal at Mackenzie and expert at organizational design, defines agility as "the ability of the organization to be dynamic, nimble, flexible, and moving fast to address change, while still focusing on core operations and key planning process." According to the author, agility can manifest itself in three key areas: Structure, where resources are allocated; governance, where decisions are made; and processes, where operations are done.

  Let us use this framework to show you how you can embed the concept of agility during the implementation of ISO 56002:2019.

  - Structure: Create a cross-functional and self-directed team to manage innovation

    Organizational structure is an important tool that helps companies deploy strategies and allocate resources, but for innovation, sources such as the functional approach and the matrix approach are not appropriate. Cross-functional team and self-managed teams are more recommended for managing innovation processes and operations because they provide better flexibility to changes and they are nimbler in addressing issues that may affect innovation. When choosing your teams, focus on diversity. The diversity of professional backgrounds, as well as cultural background, helps team members become more creative and avoid group thinking and other pitfalls to teams.
  - Governance: Review your decision-making process

    Innovation activities and procedures need a decision-making process that is fast and flexible. You cannot make decisions about creativity and innovation in the same way you make decisions about budget planning. A start-up mindset is what is needed for innovation management. Empower your team leaders and members to make decisions by allowing them room for failure. Agility requires team members to act fast, and if team leaders have to wait for decisions or have to go through different lines of bureaucracy to make decisions, then your innovation operation will suffer.

- Process: Standardize your key processes

  Companies struggling with implementing processes struggle because either the process is too complex, i.e., not intuitive, or it has not been well taught to employees. A process that is ill-designed, or not well-explained, creates bottlenecks for the workflow and adds more waste. People spend endless time in meetings asking basic questions about steps that are sometimes obvious. We hear often people saying that the process is getting in the way of innovation. This can be true only if the process is too complex for people to digest and understand. Excellent organizations invest tremendous time to create intuitive processes that are standardized so people can implement them seamlessly. The study conducted by Aaron Smith of McKinsey shows that signature processes, such as Amazon's synchronized supply chain, or P&G's product development and external communication processes, are critical elements of the companies' strategy. These companies invest tremendously in designing their processes in an intuitive way and spend much time training employees on how to implement them so people know who does what at different touch points of the workflow.

Generally speaking, there are three problems that hinder the performance of a process:

- Lack of governance: In this situation, the process is not well mapped out, or the person in charge is not identified correctly on the process map. People spend a lot of time figuring out where to go to execute simple tasks. We suggest that you spend time mapping out different workflow procedures that relate to the management of innovation initiatives, programs, and activities, and create an innovator's journey that clarifies the process of employees willing to be part of the innovation process.

- Lack of employees' skills: The second obstacle that tends to hinder the performance of a process is the lack of skills and information. A subprocess can hinder the main process if employees lack the competency they need during the execution of the process. Here you need to spend time teaching people how to use the process, and provide them with information, knowledge, and training to become nimble. Create a common language around that process so you can easily standardize it.
- Lack of employees' engagement: This situation can affect tremendously different innovation operations. In this situation, people are not "interested" in doing what they are supposed to do, even when the process is simple and intuitive. People lacking engagement start creating shortcuts, performing at the minimum threshold, not going the extra mile to fix and improve things. Your role as the innovation leader is to empower employees by creating incentives that engage them. Motivate people and make them become the center of the process so they know that if they do not do their parts, innovation operations get affected.

Clause 8.2: Managing Innovation Initiatives

Although the ISO Innovation Standards suggests that you use a single or combination of approaches in implementing your innovative system, the organizations we have worked with and for have used a customized approach driven by their current systems and different output entities.

In Subclause 8.2, ISO 56002:2019 outlines ways and approaches of implementing the innovation initiative. ISO 56002:2019 provides enough freedom and flexibility in terms of what approach or approaches you would want to use. For instance, you can implement the innovation initiative internally at the level of one or more internal units (department, business units, division group), or collaboratively with the partner with whom you developed the initiative. You can also combine both approaches to implement the initiative.

By now hopefully we all have a very good understanding of what innovation is. In Subclause 8.2, the standard has added a directional guideline by merging the word "innovative" with initiatives in suggesting that this is the way an individual or organization should manage their innovative activities.

To be able to effectively implement and use this Subclause of 56002, we have to understand what an initiative is defined as. There are many extremely different understandings and definitions of the word "initiative." It is a word that is used in many different ways to support a different meaning.

- Definition of initiative: Initiative is the power or opportunity to act or take charge before others do.
- Definition of initiative (Webster): Initiative is the ability to assess and initiate things independently. Synonyms: enterprise, resourcefulness, capability, imagination, imaginativeness, ingenuity, originality, creativity, drive, dynamism, ambition, ambitiousness.
- Definition of initiative (Cambridge): Initiative is a new plan or process to achieve something or solve a problem: Example: The peace initiative was welcomed by both sides.

Clause 8.2 is further divided into two clauses. They are:

- Subclause 8.2.1: Managing Each Innovative Initiatives
- Subclause 8.2.2: Identify How to Implement Each Innovative Initiative

Subclause 8.2.1: Managing Each Innovative Initiatives

For the sake of interpreting correctly Subclause 8.2, I believe the Cambridge dictionary's definition is closer to being in line with the activities recommended in this subclause of 56002. Most people using the standard could interpret it as programs for a project.

The purpose of Subclause 8.2.1: To provide guidance and the type of support for processes, measurement systems, and resources including availability to adequately staff with the required skills in keeping with the innovation/project plan.

Suggesting that a wide range of organizational design be considered using a single approach or combination of approaches. Consider the many options available to customize your organizational structure.

> *In an effective organization there is congruence between purpose, strategy, processes, structure, culture, and people. It is the challenge of the leaders to orchestrate this alignment and to still promote innovation and change.*
>
> **—David J. MacCoy**

Harrington points out, "A major restructuring change in any organization should only be undertaken when it will produce a very significant performance improvement and then it must be accompanied with an effective Organizational Change Management plan."

A 2016 Deloitte study found that "only 26% of large companies (> 5,000 employees) were functionally organized (i.e., sales, marketing, finance, engineering, service, etc.) and 82% were either in the process of reorganizing, planning to reorganize or had recently reorganized to be more responsive to customer needs" (Bersin Josh 2017). Overall, "92% of the companies … surveyed cited 'redesigning the way we work' as one of their key challenges, making this the #1 trend of the year."

What is driving this need to reorganize? Traditional models of organizational structure, based on functional areas, are not meeting the needs of customers or the workforce in today's environment of innovation. Information technology and instant communications have led to a shift from traditional functional and hierarchical models to network-based models that offer more rapid collaborations across more diverse populations. Effective structures must encourage such interactions, sharing of knowledge, and effective communication between customers, workforce, partners, and management.

There are at least 11 different types of organizational structures used today. Listed below are the most commonly used organizational structures. There are advantages and disadvantages to each of them, making it too risky to specify *one-size-fits-all* organizational structures.

- Functional
- Vertical
- Bureaucratic
- Decentralized
- Geography
- Product
- Customer
- Case Management System
- Process Based Network
- Matrix
- Hybrid

Here are some important additional definitions:

- Structures: The arrangement of and relations between the parts or elements of something complex. Structure is a system that outlines how organizational activities, such as task allocation, coordination, information flow, and supervision, are directed in order to achieve the goals of an organization. Structure can be considered as the viewing glass or perspective through which individuals see the organization and its environment. "Flint is extremely hard, like diamond, which has a similar structure."
- Organizational structure: A system that outlines how certain activities are directed in order to achieve the goals of an organization. These activities can include rules, roles, and responsibilities. The organizational structure also determines how information flows between levels within the company.
- Innovative: A person or thing that introduces new ideas; original and creative in thinking. Having new ideas about how something can be done. "An innovative thinker."
- Innovative organizational structure: A person or thing that introduces new ideas; original and creative in thinking through a system that outlines how certain activities are directed in order to achieve the goals of an organization.
- Hybrid: A way of structuring an organization that replaces the conventional management hierarchy. Power is distributed throughout the organization, giving individuals and teams greater autonomy to self-organize and take rapid action, while staying aligned to the organization's purpose.
- Entity: An organized array of individual elements and parts forming.

Subclause 8.2.2: Identify How to Implement Each Innovative Initiative

The purpose of Subclause 8.2.2: In Subclause 8.2.2, the primary focus is on how individual innovative initiatives are organized to accomplish the goals and objectives of maximizing the use of the resources that are available to them. This allows a great deal of latitude in selecting the initiative organizational structure and the approaches they used to meet these objectives within cost, schedule, and operational performance. For instance, you can implement the innovation initiative internally at the level of one or more internal units (department, business units, division group), or collaboratively with the partner with whom you developed the initiative. You can also combine both approaches to implement the initiative.

Although you should consider all of the organizational structure approaches available, Subclause 8.2.2 recommends that particular consideration should be given to

- cross-functional organization,
- crowd sourcing,
- clusters of organization,
- outsourcing,
- acquisitions, and
- partial spin-offs.

Although the organizational structure of the individual innovation initiatives teams is very important, what is even more important is the team talent mix. It is even more important to have a balance of technical and personal skills. People with different backgrounds, personalities, skills, and performance objectives provide the source of differing viewpoints that often create a more innovative result.

Subclause 8.2.2 does not recommend a specific organizational structure. It suggests that you consider a number of alternatives before you make a final decision, which should be made based upon optimizing the amount of resources that are used to meet the initiative's goals and objectives.

Many innovation experts believe that creativity and innovation occur in so many different ways that it is impossible to define an innovation process. But 56002 is able to define the major activities that should be considered in defining an optimum process for a creative innovative entity. This is true when you are discussing the act of creating a new and unique concept or approach. But developing the individual innovative concept is only a small part of the innovative process. Often the activities in developing and deploying the concept are more expensive from a financial and time standpoint than the identifying, creating, and validating activities. Frequently the cost of deployment of the concept is so great that innovative concepts cannot be implemented. It is for this very reason that the organization needs to be managed; each individual initiative before becomes a value drain rather than a value-added entity. Processes are the basic building block used to manage the organization's operations and the innovative process must be able to consider and manage in harmony with the other organization's processes.

(Jane Keathley, MS, PMP, made some important contributions to this discussion on Subclause 8.2.2.)

Clause 8.3: Innovation Processes

"The organization should configure the innovation processes to suit the innovation entity."

> It defines a process where an innovative opportunity can be transformed into a deployable solution. It is important to point out that this basic innovation management process will not totally incorporate all of the activities defined. Each initiative should modify the basic process to meet the unique needs of the individual entity. (©ISO. This material is excerpted from ISO 56002:2019, with permission of the American National Standards Institute (ANSI) on behalf of the International Organization for Standardization. All rights reserved.)

The *Enhance* cloud service at **EDGESoftware.cloud/managing-innovation** supports just such customization. You can start with the Best Practice processes being presented in this book and modify them to adapt to your particular needs for each project. Then you can manage the project as it progresses along your processes.

Subclause 8.3.1: IMS Overview

The IMS approaches configuration should be designed to meet the needs of individual initiatives. In a typical midsize-to-large company, the IMS flows across many smokestacks that make up the organization. Some of them are as follows:

- Marketing
- Human resources

- Research and development
- Product engineering
- Manufacturing engineering
- Industrial engineering
- Quality engineering
- Production control
- Purchasing
- Production
- Finance
- Sales
- Maintenance and repair

Yes, almost all the functions in the organization are in some way related to the IMS. To have an effective IMS, the roadblocks between these functions need to be broken down and work smoothly and flawlessly together. Fortunately, or maybe unfortunately, each of these areas has its own internal systems required to optimize its efficiency, effectiveness, and adaptability. These areas have a tendency for the individual functions to suboptimize the total system because their measurement and procedures are not in harmony with each other. I remember one instance where the product engineering group got awards for the unique creative design they came up with. Later on, we discovered that it cannot be manufactured at a price that made it a viable product.

But once again, a function that is trying to do its very best and is excelling in its performance may generate a negative impact on the total IMS. It is called suboptimization. It is the total system's performance that counts, not the performance of the individual activity. There are many difficult and challenging parts of an IMS. To get it to operate smoothly and harmoniously as a tool system is often the biggest challenge. The word "optimization" seems to be the right objective that an organization should have related to each part that impacts the total system.

- Definition of optimization: Optimization is defined as an act, process, or methodology of making something (e.g., a design, system, or decision) as fully perfect, functional, or effective as possible.

An organization's structure is directly related to its performance, and the design of the structure needs to be such that peak performance can be achieved. The influence of its structural design extends to the organization's ability to innovate and to the outcomes of each of its key performance systems. This chapter explores ways in which the structures of these key systems – management, product, process, sales and marketing, and customer service – factor into successful innovation.

One of the major benefits of management restructuring is to increase the employee's sense of ownership, dedication, and pride within the organization.

This subclause recognizes that there is no predesigned step-by-step, task by task, process that is applicable to all innovative activities. It is obvious that a continuous improvement project will cycle through a much simpler process than the process used to support a new and revolutionary product. Developing a new and creative advertising campaign process is different from designing a replacement for the 747 jetliners. Keeping all this in mind provides an overview of the typical innovation project's basic activities. Note: Improvement process basic activities are defined in Chapter 4 of this book.

BUILDING BLOCKS OF THE TIME PYRAMID

TIME uses 16 key Building Blocks (BB) to construct an organizational profile designed to consider all of the individual stakeholder's desires. These building blocks are strategically aligned with each other to increase the organization's efficiency, effectiveness, and adaptability. This combination of building blocks makes up a pyramid that is commonly known as the TIME Pyramid (see Figure 2.1).

- BB1: Value to Stakeholders (The Foundation)

  This foundation is built on bedrock to provide maximum stability to the pyramid. It provides assurance to the stakeholders that the organization's activities are stable and well-constructed. Without a good foundation, no matter how elaborate the construction is, the organization is doomed for failure. Too many of the present technologies are built on a "sand" base. As such, they look beautiful for a period of time and then slowly decay, taking the organization's culture, investors' money, and employees' jobs with it.

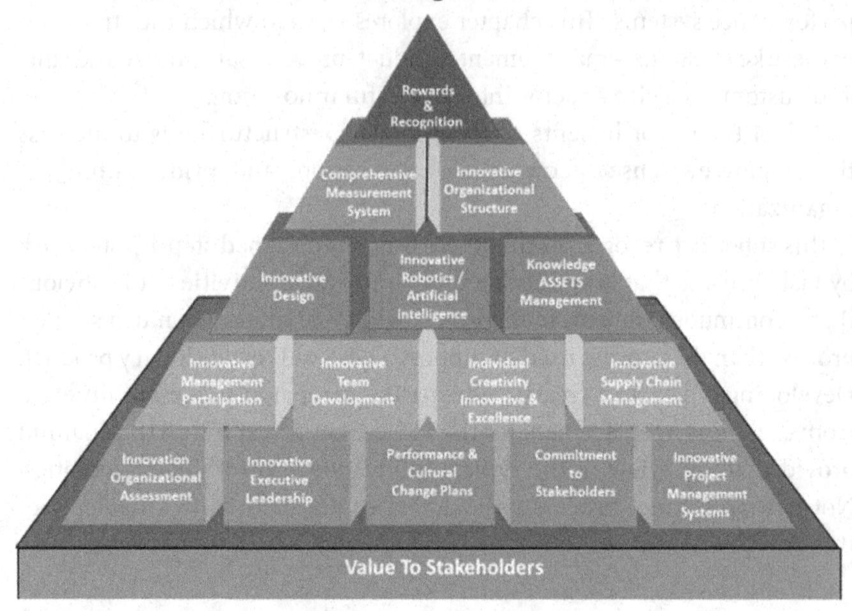

FIGURE 2.1
The TIME Pyramids for Innovation

- BB2: Innovative Organizational Assessment

  It is not practical to start any type of innovative improvement effort without establishing what your present situation is. One of the major mistakes many organizations make is thinking that the executive team has an excellent understanding of what problems the workforce is facing. We often find out that the executive team frequently has a more positive view of the organization's operations, and the employees have a very different opinion.

  It is absolutely essential that any assessment of an organization collects information related to the needs, expectations, and desires of the executive team, middle management, and the employees. Once this is done, the organization is in a position to compare the organization's strengths and weaknesses as viewed from these three separate levels. (Note: It is a rare organization where a single survey or assessment is adequate to characterize the culture of the organization and identify opportunities for major improvement.)

- BB3: Innovative Executive Leadership

If you're going to sweep the stairs, always start at the top.

—My grandmother

Top and executive management must do more than just support TIME. We like to start our innovation improvement initiative with the board of directors. The primary responsibility of the CEO of an organization is to meet the requirements of the board of directors. The performance is not based upon what they say; it's based upon what the board of directors say. Of course, the total executive team must be part of the process and participate in designing the process, assigning resources and giving freely of their personal time. The start of any improvement process is the total executive team belief and leadership to make it successful.

- BB4: Performance and Cultural Change Management Plan

All employees need to understand why the organization is in existence, what the behavioral rules are, and where the organization is going. This direction must be well-communicated to the stakeholders, and there needs to be an agreed-to plan on how the organization wants to change. That is what a business plan does for an organization. It sets the direction of the business, what products are going to be provided, what markets are going to be serviced, and what goals need to be reached in the future. Without an agreed-to, well-understood business plan that is implemented effectively, the organization has no direction, so it meets its goal of going nowhere.

A business plan setting on someone's desk is no plan at all.

—H. James Harrington

- BB5: Commitment to Stakeholders' Expectations

Every organization has an obligation to the individuals who are impacted by the organization's activities. This includes investors, management, employees, suppliers, customers, consumers, the community, interested parties, the employee's family, etc. The investor wants decreased cost so that bigger dividends can be paid. One of the biggest problems top management faces is how to balance the activities within the organization so that all the stakeholders have a win–win impression of the way the organization is managed.

- BB6: Innovative Project Management Systems

  Project Management Institute has just issued an updated version of their standard called "PMBOK." It is a well-prepared, comprehensive document that provides detailed guidance for large and small projects. The biggest innovation improvement opportunity in many companies is to address high project failure rates. To decrease the number of project failures, I recommend reading *Effective Portfolio Management Systems* (CRC Press 2015).
- BB7: Innovative Management Participation

  One of your biggest problems is the lack of applying innovation to the management methodology. This building block is designed to get all levels of management actively participating (out on the playing field) in the improvement effort. Having management feeling comfortable in a leadership role is essential to the success of the total process. It is important that you bring about the proper change in top, middle, and first-line managers and supervisors before the concepts are introduced to the employees. Most organizations have done a poor job of preparing management for their new leadership role. And, as a result, they are still managing using the same principles in the 1980s. All too often the management rule is, "Do what I say – not what I do."
- BB8: Innovative Team Development

  The organization needs to take advantage of manager and employee teams to take maximum advantage of the improvement opportunities. Everyone involved in the organization's change process is a key ingredient in today's competitive business environment. This building block develops team concepts as part of the management process and prepares all employees for participating in a team environment. One of the prime outputs from the team environment is a sense of being a member of the organization and a feeling of cooperation between the individuals within the organization. Developing a team environment will have a big impact on employee morale, efficiency, and effectiveness.

  I have often heard the theory that a team of two generates three times the output of one. I've seen many occasions when a team of two generates 0.7 times the output due to the compromises required to obtain a consensus decision.

- BB9: Individual Creativity, Innovation, and Excellence

  Management must provide the environment, as well as the tools, that will allow and encourage employees to excel, take pride in their work, and then reward them based on their accomplishments. This is another key ingredient in every winning organization's strategy. You can have a good organization using teams, but you can have a great organization only when each employee excels in all the jobs he/she is performing. Care must be taken to have a good balance between team cooperation and individuals who strive for excellence in all their endeavors. The two concepts need to work in tandem, not compete with each other. Typically the more advanced organizations will have empowered the people to take action, often eliminating the need for a team. For example, if a skid is sitting in the walkway, an empowered person will move it out of the way. On the other hand, a team would typically hold 10 meetings to do the same thing.
- BB10: Innovative Supply Chain Management

  Winning organizations have winning suppliers. The destiny of both organizations is inevitably linked. Once the innovative improvement process has started to take hold within the organization, it is time to start to work with your suppliers. The objective of this partnership is to help them improve the performance of their output and increase their profits, while reducing the cost of their product or service to you. It's a search for that win–win situation that benefits both you and your suppliers.
- BB11: Innovative Design

  An innovative design is not one that gradually changes even though the change is in a positive direction. An innovative design is one that jumps forward in the march of progress, rather than a step forward. From the consumer standpoint, an innovative design has to be one that is significantly better than any other one that's available. From the organization's standpoint, it needs to bring in more value added than the cost to develop, produce, sell, and maintain it through the warranty period. Over an (unfortunately) short time period, an innovative design can become no longer innovative as it is no longer significantly better than what is available from other sources. The first time a man cooked meat to eat over a fire was innovative; having a barbecue out in the backyard is not innovative.

After 70 years where the biggest percentage of my time was spent doing problem-solving, I realize that in most cases, poor design of products, processes, organizational structures, and methods were the real root causes of the problem, and we were focusing on correcting symptoms, rather than preventing a repeat of the same problem in the next product cycle. An innovative designer delivers a design that is efficient, effective, and adaptable in addition to meeting customer expectations. The current trend of focusing on minimizing risk needs to give way to prevent errors from occurring. To accomplish this, tools like Design for X needs to be incorporated in a design methodology and design evaluation. Typical Design for X techniques are as follows:

- Design for manufacturability
- Design for reliability
- Design for repair ability
- Design for safety
- Design for costs

In addition, increased emphasis needs to be placed on knowledge management. Most designs are reviewed by a number of functions so that the individual function is not held accountable for finding errors. All errors that occur after a design review should be charged to the organizations being paid to do the design review, not to the design department.

Unfortunately, we have all become accustomed to using our customers as the final testers. It's fast, quick, and sloppy. To offset this trend you need to improve the effectiveness of tools like Business Process Improvement, Total Quality Management (TQM), Activity-Based Costing, and Lean Six Sigma. With today's short product cycle times, it's too late to correct any problem manufacturing processes have started. By the time you find and correct a problem, the manufacturing-the-product cycle is over with and the organization is just left with an extensive recall. It means that we need to develop innovative ways to evaluate potential and improvement opportunities.

- BB12: Innovative Robotics/Artificial Intelligence

  This building block focuses on how to design and maintain product and services delivery processes so that they consistently satisfy external and internal customers and the people who consume the end product. Innovative use of technology, automation, and artificial intelligence has drastically changed the way our processes are designed and function. Automation has made concepts like Six Sigma practical in our manufacturing processes. Technology provides us with new products almost on a monthly basis. Artificial intelligence provides the capacity for a computer to perform operations analogous to learning and decision-making by humans, using an expert system, a program for CAD or CAM, or a program for the perception and recognition of shapes in computer vision systems. In many applications, it is impossible for humans to make decisions as fast or as correctly as artificial intelligence can. The combination of innovative personnel using technology, automation, and artificial intelligence is bringing us closer and closer every day to the ultimate factory of the future. Future accuracy, repeatability, dependability, and precision will not be in the hands of a human, but in the programming designed into the new computerized environment.

In BB12, we present how you can use automation, technology, and artificial intelligence to reduce costs, assist in creating new products, reduce cycle time, while improving the quality of the delivered product. Who would've believed just a few years ago that computers would be taking our order in a restaurant rather than a waiter/waitress?

- BB13: Knowledge Assets Management

Today more than ever before, knowledge is the key to organizational success. Instead of having one or two sources of information, the Internet provides us with hundreds, if not thousands, of inputs, all of which must be researched for the key nuggets of information. We are so overwhelmed with so much information that we don't have time to absorb it, so we depend on the computer to do it for us.

To make matters worse, most of the organization's knowledge is still undocumented and rests in the mind and experiences of its employees. This knowledge disappears from the organization's knowledge base whenever an individual leaves an assignment.

An organization's first challenge is how to collect the undocumented knowledge that rests in the minds of its employees. A second challenge is, "How do you prevent outside sources, including your competition, from hacking into your knowledge base?"

Because Knowledge Assets Management is such a critical part of the innovative cycle, we suggest that you read *Knowledge Management*

Excellence: The Art of Excelling in Knowledge Management (published by Payton Press, author H. James Harrington).

The *Enhance* cloud service at **EDGESoftware.cloud/managing-innovation** offers "process based Knowledge Management" for your knowledge assets. This approach allows you to attach your own knowledge assets to your custom process models so you can "access the right knowledge at the right place at the right time."

- BB14: Comprehensive Measurement Systems

  This building block helps the organization develop a balanced measurement system that demonstrates how interactive measurements like quality, productivity, and profit can either detract from or complement each other. This is important because organizations today need to consider all stakeholders, not just one of them. A change that has a positive impact on one stakeholder can have a very negative impact on another. For example, a drug that relieves back pain might also cause heart attacks.

  Only when the improvement process documents positive measurable results can we expect management to embrace the methodology as a way of life. A good measurement plan converts the skeptic into a disciple.

- BB15: Innovative Organizational Structure

  In our new environment, employees are empowered to do their jobs and are held accountable for their actions. With these changes, large organizations operating as a monolith need to give way to small business units that can react quickly and effectively to changing customer requirements and the changing business environment. This building block helps an organization develop an organizational structure that meets today's needs and tomorrow's challenges. For example, we worked with one organization where we were able to reduce two layers of management and eliminate 32 management positions. Establishing a dual ladder for management and technical people that was equivalent financially and in stature was a key activity in positioning the 32 managers whose jobs were eliminated.

- BB16: Rewards and Recognition

  The Rewards and Recognition process should be designed to pull together the total pyramid. It needs to reinforce and reward individuals that perform in keeping with the organization's desired behaviors. It also needs to be very comprehensive, for everyone hears

"Thank You" in a different way. If you want everyone to take an active role in your improvement process, you must be able to thank each individual in a way that is meaningful to him or her. There is a time for a "pat on the back" and a time for a "pat on the wallet." Your rewards and recognition process should include both.

Is the TIME methodology the only way to improve the quality of life in the United States? No, but it is one way that we believe progress can be made to make our products and services maintain or improve their position in the international market. It will require many different supporting activities to bring about the required changes that are facing us in an artificial intelligence and robotics world.

There is no better time than today for each of us to write on our forehead, "We need to be more creative." Managers need to lead the way with some staggering changes to our personnel systems. *Fortune* magazine each year publishes a list of the 100 best companies to work for and a list of the most innovative companies. If your company wasn't in the top 10 in both of these lists, there is an obvious opportunity for you to be more innovative.

THE ADVANTAGES OF A GOOD IMS

There are many advantages of having an effective IMS that meets or exceeds the relevant suggestions defined in the ISO 56000:2020 series. Some of these advantages are listed here:

- Management's expectations are clearly communicated to the employees.
- The organization performs much more predictably.
- There is international acceptance of the organization's innovation system.
- It provides a base for all the organization's improvement activities.
- It increases market share.
- It may be required by some organizations as part of their contract with their suppliers or subcontractors.
- It saves time because key procedures are documented, eliminating the need to reinvent the wheel over again each time.

- It provides a base that ensures that improvement gains are captured and internalized.
- It defines potential improvement opportunities.
- It focuses on expanding resources on items that drive added value.
- It triggers the new product development release cycle.
- It establishes a knowledge-based design environment.

We like to think of a good IMS as the stable base that other improvement efforts should be built upon. Improvements that are made in a poorly defined system are a lot like pushing a big round boulder up a steep incline. The moment you relax and turn your back on the boulder, it rolls right back down the hill, often crushing the people that have worked so hard to push it up the incline. This is what has happened to many organizations when they tried to implement process reengineering, quality circles, Total Quality Management, Six Sigma, Lean, or Activity-Based Costing. A good IMS provides the block under the boulder that keeps it from rolling back down the hill (see Figure 2.2).

Be careful about using projected performance improvement numbers because we know of no statistically sound, reasonably accurate data available that proves that installing the new IMS improved the organizations' performance. The studies we have seen did not include

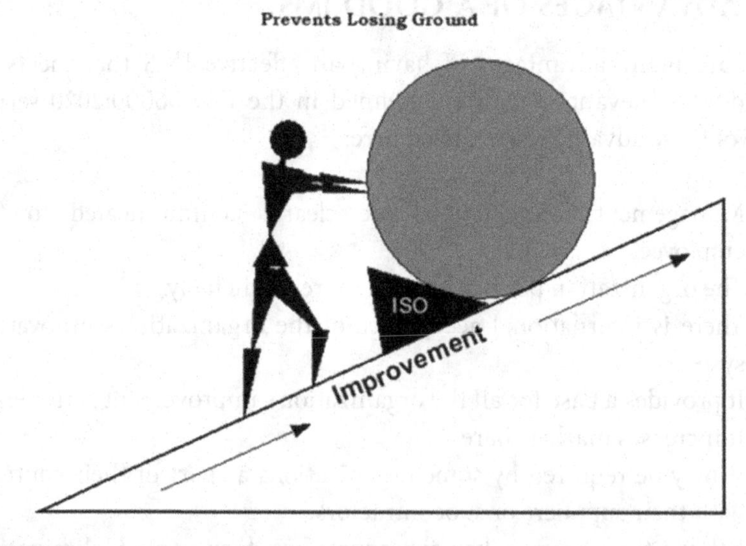

FIGURE 2.2
Maintaining Advancements in IMS

data related to the organizations' performance before ISO 56002:2019 was installed. In addition, the degrees of installation of the best practices listed in ISO 56002:2019 varies from organization to organization because ISO 56002:2019 is a guidance document, not a requirements document. To be absolutely truthful, we know of no organization that has implemented all the suggestions given in ISO 56002:2019, and we know of some organizations, rated in the top five innovative companies in the United States, that disagree with some of the suggestions. We personally believe that all of the suggestions given in the document are things that should be considered when designing your IMS, but not automatically, blindly built into your IMS.

A WORD OF CAUTION

Don't undertake the upgrading of IMS to meet the ISO 56000:2020 series requirements lightly. It will require a major commitment of the executive team's personal time and the organization's resources. In a freewheeling, entrepreneurial-type organization, defining, documenting, and implementing procedures that define how the processes operate often directly oppose the current culture and force the organization to change in ways that decrease creativity. Before you decide to upgrade your IMS, ask yourself the following questions:

- Why should the organization upgrade its innovation system?
- What benefits will the organization receive from a better innovation system?
- How will we measure these benefits?
- How much will it cost the organization?
- What will be the organization's return on investment?

IMS SUMMARY

There is no doubt about it. Innovation Management Systems are here to stay. Soon it will be a major consideration in the selection of new subcontractors. There are two prime reasons that are driving the ISO 56002:2019 landslide. The first and primary reason is that customers are demanding and expecting their suppliers and subcontractors to be innovative. The second reason for implementing formal IMS is that organizations around the world are using it to create a competitive advantage for their organization. You have two options, "You can be the bandleader out in front of the parade or the street sweeper behind the parade that's picking up the horse droppings. It's up to you."

The authors of this book, after carefully examining the pros and cons of a formal documented IMS, strongly support the concept and encourage all organizations to place a high priority on this activity within their organization's business plan. The truth of the matter is, we find it hard to understand how large- to medium-size modern organizations can operate without a formally documented IMS. Moreover, we are surprised that this issue was not highlighted and addressed back in the 2010s when innovative organizations were capturing the lion's share of the market, making it an executive issue.

3

Assessing the IMS: A Discussion of 56004:2019

ASSESSMENT APPROACHES

The Standard 56004:2019 recommends that before starting an assessment of the IMS, you need to define the approach and the framework of the assessment. In other words, you need to clarify what you are assessing, for what purpose you are assessing, and how you will conduct the major assessment steps. This includes a clear definition of the criteria you will use to evaluate the performance of innovation as well as the format and the style of the assessment report.

The Standard 56004:2019 in its Section 5.2 Understanding Different Approaches to Innovation Management Assessment provides a detailed description of the framework of the Innovation Management Assessment (IMA). This framework is made of the following components:

- Assessment objectives
- The breadth and extent of the assessment
- Assessment focus
- Expertise involved in assessment
- Data collection and data collection tools
- Data type
- Reference and comparison
- Data interpretation
- Innovation management outputs, format, and reports

DOI: 10.4324/b22993-3

Assessment Objectives

The first important component of the assessment framework is to spell out the goals and the objectives of the assessment, i.e., why do you need to conduct an assessment? You can assess your innovation management system for any reason you want to. Remember this standard is just a guidance and does not require a specific purpose. However, the standard suggests three reasons that can justify an assessment system:

- Compliance with defined target: If you have identified a specific target at the beginning of the year, it is now time to assess your performance against these targets. Targets can be tangible, such as achieving certain financial results or market position. Targets can also be intangible, like building an innovation capability, an R&D capability, or a reputation in the market of being the most "innovative company."
- Value creation from enhanced innovation management: The second reason for an Innovation Management Assessment is to check how enhancements to your innovation management system have contributed to creating and realizing value. An important dimension here is that you need to spell out the previous weaknesses, the improvements you made, and the result achieved. The goal of the assessment is to link these three dimensions in order to show the new and improved value realization and how these improvements have contributed to a better position.

  An example of an improvement of the innovation system is the introduction and the implementation of a new idea management platform that enhances the way you track the ideation process. Another example would be the integration of a supplier into the IMS and the impact of this integration optimizing resources. And a final example of IMS enhancement that might need to be assessed is the acquisition of a new distribution channel that adds a new innovation capability to the organization.
- Innovation management capabilities improvement: Another reason mentioned by the standard that helps you justify the purpose of an assessment system is the improvement you bring to innovation capability. Capability improvement is a critical dimension of an innovation management system, and it is what makes an organization a unique one. When you are able to bundle your resources with

your ability to manage innovation activities and programs, you have created a capability that is hard to copy. However, and because capabilities evolve overtime, as the organization goes through change and transformation, reassessing them becomes crucial to the enhancement of the system. Examples of capabilities that you may want to assess would be a new methodology you implemented, an acquired business unit that brought new practices, or a new marketing channel that you added to your IMS.

The Breath and the Extent of the Assessment

The second question you need to address in deciding about the IMA is the level of depth of the assessment. You can assess the entire organization, focus on one single unit of the organization, or assess a few units together. A critical dimension that you need to keep in mind is that when choosing to focus on a few units, choose units that are linked to each other so you are able to identify the interdependencies that ought to exist as well as the iteration that needs to happen in typical innovation activities. For instance, you cannot choose to assess the unit in charge of managing innovation, without assessing customer service or marketing as these two units are part of the innovation ecosystem of the organization. In my view, when designing an assessment strategy, it is better that you invest time to assess the entire organization rather than segment the organization. A comprehensive approach yields always a better understanding and fits the nature of creativity and innovation.

Assessment Focus

The next question that you need to address is deciding about the object of the assessment. In other words, what will be the focus or the object of the assessment? Will the assessment focus on strategy? Will it address the process? Are you assessing the ideation? Or the supporting technology? Common assessed objects are the suggestion system, the tools and methods supporting the IMS, and customer feedback. You can also assess supporting events such as hackathons, design thinking workshops, and brainstorming sessions. Keep in mind that when focusing on a single element, while this may allow you to understand its strength and weakness, it provides little understanding of how the system interacts with other parts of the IMS.

Expertise Involved in Assessment

The view of the expert involved in assessing your IMS is important. Beyond the report written by the expert, you also get a perspective and an interpretation of how your system functions overall. However, the expert's view could be also biased. If you choose to tap on your internal expertise to assess the innovation system, it is possible to do so using ISO innovation series documentation, guidance, and framework that are publicly available. But keep in mind that an internal expert may overlook certain dimensions that are critical to the IMS. An internal assessor may also be biased in his or her judgment. A common bias found in internal assessment is groupthink bias, in which the assessor thinks in a similar pattern as the team being assessed; the Abilene paradox, in which the assessor behaves in a way that pleases the unit being assessed most often because of the presence of someone with expert knowledge; or simply the confirmation bias, in which the assessor attempts to confirm things that he or she already knows about the unit being assessed.

An external assessor, on the other hand, while expensive, is more suited for innovation assessment. Removed from the internal politics of the organization, an external assessor may provide a fresh look at how the innovation management system functions, providing an objective view of the IMS's shortcoming and helping the organization learn new best practices. My suggestion is that you conduct an internal assessment once or twice a year, and conduct an external assessment every other year depending on the specifics of your organization, and the context in which you operate.

Data Collection and Data Collection Tools

The Standard 56004:2019 provides three different options for data collection. You can use them simultaneously, separately, or combine both of them. For instance, you may collect data using an online survey and desk research, while adding a face-to-face component like an interview or focus group. Interviews, in my view, provide more credibility and validity to the data at hand. In my experience, I always find interviews more revealing and insightful as they tell about the true strength and weakness of the organization, while also getting an insider view on the human capital of the organization. So even when you are collecting data through desk research and online surveys using automated tools, it is a good practice to add a human dimension to data collection by conducting direct interviews or focus groups with people and employees engaged with the innovation system.

Data Type

In research methods, we make generally a distinction between the qualitative approach and the quantitative approach. While in the quantitative approach we use hard data as a measurement tool to get an indication of the trends and where the organization stands in terms of objectives, target, and benchmarks, in the qualitative approach, we tend to dive deeper to get insights on the true meaning of the data. A good assessment philosophy is one that combines both approaches to develop a holistic approach to assessment. This is why I suggest that a good assessment is one that uses both types of data collection in order to help the team develop a better sense of the issues.

Reference and Comparison

When assessing an IMS, it is important that the assessment provides a reference and a comparative framework that point to the differences in performance between your system and the chosen benchmark. The standard provides three options you can use in order to create this comparative framework. The first one is the before/after option. The before/after option is a side-by-side comparison of the previous performance and the now performance in which issues are easily recognized. In this option, you show how the system used to perform and how it is performing now. The good thing about this option is that it is simple to understand and can have a visual impact on those involved in managing the system.

The second option is the comparison between actual performance versus target performance. This is also a good comparison as it shows the level of performance and the stretch that you have to engage in to improve the system. The third option is to conduct a correlation analysis to show the connections between two or more variables. This is usually helpful when you introduce a new variable such as a new capability or a new resource, and you want to show the level of impact of this new variable on the innovation system.

Finally, you can also use best practices or benchmarking. Best practices and benchmarking is a comparative framework that compares the performance of your IMS with outside systems. This is actually what I recommend. Benchmarking is helpful because it allows you to stack up the performance of your innovation system against that of your competitors.

It also allows you to learn new best practices that you can deploy in your organization. Benchmarking is also important to the innovation system because it links your innovation activities to the overall innovation ecosystem of the economic cluster in which you operate.

Data Interpretation

Data interpretation is a critical component of the overall IMA. Without a good interpretation of the connections between different variables and how the system performs overall, there is no meaning to the assessment. More importantly, data interpretation helps top leadership make decisions to adjust the system by diverting resources or changing the course of actions. Data interpretation also helps the innovation team develop a comprehensive view of how different parts of IMS are connected and how they affect each other. Finally, the data interpretation helps you formulate the recommendations needed to improve the system.

Innovation Management Outputs, Formats, and Reports

The output of the IMS is a report that helps top leadership understand the strengths and weaknesses of the system, while identifying the gaps and the ways the organization needs to improve. This is where you can tell top leadership about how you feel about the organization and what needs to be done to change or improve current performance. A good report is a report that is written in a positive tone, identifying the true strength of the team being assessed, but also allowing the organization to stretch. If the assessment is being done because of a competition like assessing the organization for a national award, then paying careful attention to the writing style and the suggestions is critical to the credibility of the report. In this type of reports, showing the weakness with evidence and pointing to the improvement and the best practice you see fit can help the organization secure the buy-ins of top leadership and make your suggestions more acceptable to lower and middle management.

Another equally important dimension of the report is data visualization. Data visualization helps the reader of your recommendations understand the report. This is also very helpful to top leadership who may not have enough patience for narratives. A visualization that describes the connections, shows the cause and effect, and describes correlations is much easier to understand (see Table 3.1).

TABLE 3.1

IMA Checklist Based on 56004:2019

Questions to Ask	Explanations
Assessment objectives	Why do we need an assessment? Is it for measuring how well we achieved our goals? Is it to add a better value? Or is it to improve an innovation capability?
The extent of assessment	What are we assessing? The entire organization, a single business unit, or the whole enterprise system?
Assessment objects	What component of the business are we assessing? Are we assessing only one dimension of the business such as strategy, execution, or ideation process? Or all the dimensions of the business?
Expertise involved	Are we conducting an internal assessment using our internal assessors? Or are we hiring an outside assessment organization?
Data collection	How is data collected? Is it through interviews, online surveys, or desk research?
Tools for data collection	How are we collecting the data? Are we collecting it manually or using an automated tool, or both?
Data type	What type of data are we using to assess innovation? Is it only quantitative data, such as financial and accounting ratios and marketing returns? Or are we also capturing qualitative data, such as people's ability to change and employees' perception of innovation?
Reference type	What references and benchmarks are we using to assess innovation? Is it a before/after and pre/post approach? Actual versus target? Or simply benchmarking best practices?
Comparison type	What comparative framework are we using? Are we assessing innovation based on previous innovation assessments conducted in the past, correlation analysis between different variables, or a simple benchmark of industry leaders' practices?
Data interpretation	How are we interpreting the data we are getting from the assessment? Are we using a normative data approach, like using a baseline measurement obtained from a large sample? Or simply interpreting the data independently from any score or measures obtained in the past?
Innovation Management Assessment output	What is the format of the report that we will generate? Is it an executive summary of the findings? A comprehensive report? Or a report that will be supported by other tools such as software and Excel sheets?
IMA recommendations	What is the purpose of IMA recommendations? Is it to enhance the innovation management system? To enhance the assessment process? Or to enhance the innovation capability of the entire organization?

SECTION 5.2.1: PERFORMANCE CRITERIA FOR INNOVATION MANAGEMENT

This section defines the components of the IMS that needs to be assessed. The components are the same as the ones used in 56002:2019 and are innovation strategy, innovation structure, innovation culture, innovation process, innovation enablers, and innovation results. Let us look at these components one by one:

1. Innovation strategy

   In 56002:2019, innovation strategy is listed under Section 5 and has two important components: leadership and strategy.

   There are a number of critical elements that you need to look at. Some of them are:

   - Leadership commitment to innovation
   - Innovation vision
   - Innovation strategy organization goals and responsibilities

1.1. Strategy

   The purpose of an innovation strategy is to identify leadership future aspirations and provide directions on where the organization needs to go while spelling out the rationale of critical choices made. Strategy helps stakeholders understand the intent of the organization. An innovation strategy should also use a specific language that sets the organization apart from the competition and helps employees focus on the goals of the strategy by providing them with a sense of clarity that helps them understand leadership decisions.

   There are four components that you need to assess:

   - Activities: Does the innovation strategy describe the innovation activities that top leadership believe are critical to the success?
   - Flexibility: Is the strategy written in a flexible manner that allows for unforeseen events to take place and to help the organization change and adapt and still perform? When looking at this component, it is important to analyze the choices of the vocabulary and the language used to develop a sense of how flexible the strategy is. Ask about the frequency of adaptation and "what-if" scenarios.
   - Communication: An important component of an innovation strategy is how well it is communicated and understood by

employees. A good innovation strategy that is well crafted but no one knows about it does not allow for the organization to create a harmony between where it wants to go, what it does, and what it says. So during the assessment, asking about different ways of how this strategy is communicated is important. More interestingly, checking whether employees know about the innovation strategy provides you with clear evidence of how and whether the strategy is well communicated and understood.

- Documentation: The final question you need to ask about the strategy is whether the document containing the strategy subscribes to document control quality procedures. For instance, is the strategy documented? And is the document containing the strategy approved, reviewed, and adapted?

Finally, and while it is understood that the focus of innovation strategy is to realize value, it is critical that you check the assumptions under which the strategy was made. The Standard 56002 suggests that a strategy is only good as long as it carries out a balance between facts, evidences, and assumptions. In other words, for each strategic decision, there must be a set of facts (such as growth in a specific sector or market demand), and a set of assumptions, such as the acquisition of a new capability that will help the organization achieve those goals. Going after those intricacies between facts and assumptions helps you objectively assess the strategy and identify the gaps that may help leaders understand how to improve their strategy during the next revision cycle of the following three standards.

- ISO 56000 Innovation Management: This document provides the vocabulary, fundamental concepts, and principles of innovation management and its systematic implementation.
- ISO 56002:2019 Innovation Management: Innovation Management System – Guidance: This document provides guidance for the establishment, implementation, maintenance, and continual improvement of an innovation management system for use in all established organizations.
- ISO TR 56004:2019 Innovation Management Assessment – Guidance: This document provides guidance for organizations to plan, implement, and follow up on an Innovation Management Assessment.

The following are soon-to-be-released documents:

- ISO 56003 Innovation Management – Tools and Methods for Innovation Partnership – Guidance
- ISO 56005 Innovation Management – Tools and Methods for Intellectual Property Management – Guidance
- ISO 56006 Innovation Management – Tools and Methods for Strategic Intelligence Management – Guidance
- ISO/DIS 56007 – Tools and methods for idea management - Guidance
- ISO 56008 Innovation Management – Tools and methods for innovation operation measurements – Guidance

4

The PIC Cast Members

INTRODUCTION

The purpose of this chapter is to alert the reader what type of people will be involved in the assessment and what levels of expertise they should have. A product innovation system assessment involves most of the players within the organization and, as a result, must be carefully planned and executed to minimize the impact on normal work activities.

The IMS is made up of many different and interrelated processes that need to be evaluated during this assessment. This chapter points out key processes that need to be considered. Processes vary from organization to organization so the list provided is much more detailed than most organization are using.

THE CAST OF PLAYERS

Before we delve into the individual activities that that make up the Project Innovation Cycle (PIC), we believe that it would be worthwhile to introduce you to each of the major players and describe their normal responsibilities and accountability. Many functions/individuals are typically involved in the processing of an innovative project through the IMS. Some of the key players are:

- Research and development*
- Product engineering*
- Marketing*

DOI: 10.4324/b22993-4

- Sales*
- Finance
- Project management*
- Personnel
- Production control
- Manufacturing engineering*
- Testing engineering
- Industrial engineering
- Quality assurance
- Legal
- Procurement
- Executive sponsor*
- Project
- Project leader*
- Other minor players

Note: The organizations with an * after their name are organizations that are deeply involved throughout the entire assessment process. The other organizations are primarily involved when the evaluation is directly associated with their area of responsibility.

Although we plan on addressing the activities of the major innovation players, Table 4.1 is a list of the major processes' variance functions that would be assigned to manage.

TABLE 4.1

Typical Business Process List

Function	Process Name
Development	Records Management
	Acoustics Control Design
	Advanced Communication Development
	Cable Component Design
	Reliability Management
	Cost Target
	Design Test
	Design/Material Review
	Document Review
	High Level Design Specification
	Industrial Design
	Inter-Divisional Liaison

(Continued)

TABLE 4.1 (CONTINUED)

Typical Business Process List

Function	Process Name
	Logic Design and Verification
	Component Qualification
	Power System Design
	Product Management
	Product Publication
	Release
	System Level Product Design
	System Reliability and Serviceability
	System Requirements
	Tool Design
	User/System Interface Design
	Competitive Analysis
	Design Systems Support
	Engineering Operations
	Information Development
	Interconnect Planning
	Interconnect Product Development
	Physical Design Tools
	Systems Design
	Engineering Change Management
	Product Development
	Tool Development
	Development Process Control
	Electronic Development
	Phase 0/Requirement
Distribution	Receiving
	Shipping
	Storage
	Field Services/Support
	Teleprocessing and Control
	Parts Expediting
	Power Vehicles
	Salvage
	Transportation
	Production Receipts
	Disbursement
	Inventory Management
	Physical Inventory Management

(*Continued*)

TABLE 4.1 (CONTINUED)

Typical Business Process List

Function	Process Name
Financial Accounting	Ledger Control
	Financial Control
	Payroll
	Taxes
	Transfer Pricing
	Accounts Receivable
	Accrual Accounting
	Revenue Accounting
	Accounts Payable
	Cash Control
	Employee Expense Account
	Fixed-Asset Control
	Labor Distribution
	Cost Accounting
	Financial Application
	Fixed Assets/Appropriation
	Intercompany Accounting/Billing
	Inventory Control
	Procurement Support
	Financial Control
Financial Planning	Appropriation Control
	Budget Control
	Cost Estimating
	Financial Planning
	Transfer Pricing
	Inventory Control
	Business Planning
	Contract Management
	Financial Outlook
Information Systems	Applications Development Methodology
	Systems Management Controls
	Service Level Assessment
Production Control	Consignment Process
	Customer Order Services Management
	Early Mfg. Involvement and Product Release
	EC Implementation
	Field Parts Support

(*Continued*)

TABLE 4.1 (CONTINUED)

Typical Business Process List

Function	Process Name
	Parts Planning and Ordering
	Planning and Scheduling Management
	Plant Business Vols. Perf. Management
	Site Sensitive Parts
	Systems WIP Management
	Allocation
	Inventory Projection
	New Product Planning
	WIP Accuracy
	Base Plan Commit
	Manufacturing Process Record
Purchasing	Alteration/Cancellation
	Expediting
	Invoice/Payment
	Supplier Selection
	Cost
	Delivery
	Quality
	Supplier Relations
	Contracts
	Lab Procurement
	Non-production Orders
	Production Orders
	Supplier Payment
	Process Interplant Transfer
Personnel	Benefits
	Compensation
	Employee Relations
	Employment
	Equal Opportunity
	Executive Resources
	Management Development
	Medical
	Personnel Research
	Personnel Services
	Placement
	Records
	Suggestions

(*Continued*)

TABLE 4.1 (CONTINUED)

Typical Business Process List

Function	Process Name
	Management Development/Research
	Personnel Programs
	Personnel Assessment
	Resource Management
Programming	Distributed Systems Products
	Programming Center
	Software Development
	Software Engineering
	Software Manufacturing Products
Quality	New Products Qualification
	Supplier Quality
Site Services	Facilities Change Request
Miscellaneous	Cost of Box Manufacturing Quality
	Service Cost Estimating
	Site Planning

Research and Development

The belief that in innovation there's nothing wrong with failing/making an error because you learn from your failures probably started in R&D. The very nature of R&D is to look at the many combinations to find the one that best addresses the opportunity. Basically, each individual evaluation goal is designed to answer the question, "Is this the right or the wrong way to address the opportunity?" If you can complete the evaluation with a "yes or no" answer, it is a successful evaluation. A field evaluation is one that you cannot draw such a conclusion from.

R&D is separate from most operational activities performed by a corporation. The research and development is typically not performed with the expectation of immediate profit. Instead, it is expected to contribute to the long-term performance of the organization.

R&D includes activities that companies undertake to innovate and introduce new products and services. It is often the first stage in the development process. The goal is typically to take new products and services to market and add to the company's bottom line.

The R&D department is responsible for innovation in design, products, and style. It is responsible for creating new products, improving existing consumer products, and exploring new ways of producing them. R&D will provide advancements in products that range from simple evolutionary to addressing the discovery of new concepts.

They are responsible for the following three types of development activities:

1. Basic research: Directed at developing knowledge not related to a specific process or product. Its spending accounts for 0.5% of the GDP.
2. Applied research: Research directly focused on specific opportunities and customer needs. Its spending accounts for 0.5% of the GDP.
3. Developed research: Activities to help bring a specific opportunity into production. Its spending accounts for 1.8% of the GDP.

The manufacturing sector accounts for about 70% of all industry research and development expenditure, and most of that is devoted to computer and electronic product inductors and the chemical industry.

Basic research is aimed at a fuller, more complete understanding of the fundamental aspects of a concept or phenomenon. This understanding is generally the first step in R&D. These activities provide a basis of information without directed applications toward products, policies, or operational processes. Basic research helps the company acquire new knowledge but doesn't have any specific application or use in mind. Think of it as research for the sake of research. It is most often performed by the government or universities. Most private industries cannot afford to invest in basic research.

Applied research entails the activities used to gain knowledge with a specific goal in mind. The activities may be to determine and develop new products, policies, or operational processes. While basic research is time-consuming, applied research is painstaking and more costly because of its detailed and complex nature. Applied research is also done to acquire knowledge; it's done with a specific goal, use, or product in mind – such as how to build a better mousetrap.

Development is when findings of a research are utilized for the production of specific products, including materials, systems, and methods. Design and development of prototypes and processes are part of this responsibility. A vital differentiation at this point is between development and engineering and manufacturing. Development is research that generates required knowledge and is designed for production and converts these into prototypes. Engineering is the utilization of plans and research to produce commercial products.

R&D's typical output is a design concept for an improved current product or process or a new concept that is transferred to product engineering for refinement that is transferred to product engineering for implementation. Even in these conditions, R&D still is responsible for working with the innovation team to develop the entity.

Product Engineering

The product engineering department is responsible for transforming the concepts developed in R&D and transforming them into procedures that can be used to apply them to the manufacturing process. It includes not only product documentation but also the paperwork required to direct and control the compliments and special processes.

A product engineer is responsible for the design and creation of products that perform constant marketing analysis of competing products to determine established customer needs and requirements.

Product engineering takes care of the entire product lifecycle starting with the development of product engineering control documents and specifications. It continues to assist manufacturing in the problem analysis activities that result in changes to the engineering documents. Test engineering may be a part of manufacturing engineering dependent upon the complexity of the testing activities.

What are the various phases of product engineering? Various phases of product engineering are:

- Product Ideation
- Product Architecture
- Product Design
- Product Testing
- Product Migration and Porting
- Technical Support
- Sustaining Engineering

The product engineering department specializes in designing, building, and testing the prototype of a fabricated product and controlling subsequent changes in the construction and material of the product. They transform the R&D concept into the customer-shippable product, which includes the production process development and implementation. It is the halfway house between R&D and manufacturing. Recently there has been an increased focus on automation and applications of artificial intelligence to the manufacturing system.

Product engineering usually entails activity dealing with issues of cost, producibility, quality, performance, reliability, serviceability, intended lifespan, and user features. It includes design, development, and transitioning to the manufacturing of the product. The term encompasses developing the concept of the product and the design and development of its mechanical, electronics, and software components. The product engineer is responsible for:

- Developing product ideas based upon customer needs and expectations
- Managing the budget-risk requirement methods for projects
- Coordinating prototype
- Overseeing the mass production of prototypes
- Identifying customer problems and initiating corrective action
- Originating production level engineering documents
- Performing market analysis of competing products
- Analyzing market and industry trends and conditions
- Analyzing feasibility, cost, and return on investment for product ideas and changes

Manufacturing Engineering

The manufacturing engineering department is responsible for developing the processes, workflow, assembly procedures, test procedures, equipment requirements, supplier qualification, etc. that have an impact on the ability to take the engineering design documentation and produce a high-volume product that meets or exceeds the engineering spec performance requirements. Manufacturing engineering usually entails activity dealing with issues of cost, producibility, quality, performance, reliability, serviceability, intended lifespan, and user features. These product characteristics are generally all sought in the attempt to make the resulting product attractive to its intended market and a successful contributor to the business of the organization that intends to offer the product to that market.

The manufacturing engineering department is responsible for:

- Defining the yield road map and driving the fulfillment during ramp-up and volume production
- Identifying and realizing measures for yield improvement, test optimization, and product cost-ability methods
- Defining qualification plan and performing feasibility analysis
- Setting up that data collection and analysis systems
- Presenting corrective action on problems and potential problems

Although a manufacturing engineer is primarily faced with technical problems/developments, he also has to have a high degree of social skills. Manufacturing engineers are almost always part of the team or the leader of the team that requires leadership skills.

Manufacturing engineers are the technical interface between the component development team and the production side (front-end and back-end), especially after the development phase and qualifications when the high-volume production is running.

Manufacturing engineers improve the product quality and secure the product reliability by balancing the cost of tests and tests coverage that could impact the production fall-off. They support failure analysis requests from customers.

Marketing

The difference between marketing and sales lies in how close you are to converting a potential customer to an actual customer.

—Laura Lake

Dr. Philip Kotler defines marketing as

> the science and art of exploring, creating, and delivering value to satisfy the needs of a target market at a profit. Marketing identifies unfulfilled needs and desires. It defines, measures and quantifies the size of the identified market and the profit potential. It pinpoints which segments the company is capable of serving best and it designs and promotes the appropriate products and services.

Marketing is responsible for reaching out to new potential customers and generating interest in the selling organization's business. It requires information and knowledge that allows your organization to understand present and potential customers' needs and expectations. It gathers information by reviewing published market research reports, asking for sales team feedback, or carrying out a survey using a market research firm.

Marketing should also monitor product review sites and social media, such as Facebook and Twitter. It should collect knowledge and information on consumers' needs and attitudes toward products wherever they can find it.

Your marketing function should be involved in the following seven different types of activities.

1. Promotion

   This activity is designed to make customers and prospects aware of your products and your company. Typical activities included in the promotion are:

   - Advertising
   - Direct marketing
   - Telemarketing
   - Public relations

   (Note: You can easily see why this is so important in the IPC.)

2. Distribution

   Your distribution strategy determines how and where customers can obtain your products.

3. Product/Service Management

   The information collected identifies the features to incorporate in new products and product upgrades.

4. Selling

   Marketing and selling are complementary functions. Marketing creates awareness and builds preference for a product, helping company sales representatives or retail sales staff sell more of a product. Marketing generates leads for the sales team. It also creates interest and awareness in the product being sold. It is estimated that over 70% of the sales process is complete before a potential customer talks to a salesman.

5. Pricing

   The price of your product is not based upon what it costs to deliver it but what the market sets the price at. Marketing needs to look at all the factors in the customer's environment as well as the pricing of the competitors in order to recommend a pricing strategy for a new or upgraded product or service. Pricing plays a key role in establishing your production demands and organizational profitability.

6. Forecasting

   Marketing is responsible for providing estimates on quantity and sales price. For potentially new products, their estimates frequently make the difference between the item being included in the organization's portfolio or being dropped from consideration. Three-month sales forecasting that is higher than demand results in large inventories being built up, or forecasting that is lower than demand results in loss of sales.

7. Financing

   Marketing programs that strengthen customer loyalty help to secure long-term revenue. Financing also plays a role in marketing's success by offering customers alternative methods of payment, such as loans, extended credit terms, or leasing.

(Source: Greenville Technical College book on project management training, page 294)

Executive Project Sponsor

- Definition: Project sponsor is an individual in an organization whose support and approval are required for a project to continue.

The project sponsor is usually the "owner" of the project; he or she provides overall project leadership and should be a vocal advocate of the project. The project sponsor generally has the delegated authority of the steering committee to assist with business management and project management issues that arise outside of the formal charter of the steering committee. This individual is ultimately accountable for ensuring that business benefits are identified, validated, and delivered in a timely manner, and he or she should lead discussions with steering committee members on project issues and proposed changes to project scope.

The project sponsor should be actively engaged in the project and readily available to the project manager for making project-related critical decisions. As the individual who leads discussions on the project, the project sponsor will usually chair project discussions and make the final decisions.

In addition, the project sponsor has several other important responsibilities. For example, he or she will help develop and sustain project focus, help attain project resources, and resolve issues that represent major barriers that impede success. Some of the responsibilities of the project sponsor include, but are not limited to, the following:

- Understanding the business need
- Initiating project requests
- Determining project priority against other project requests
- Evaluating the project business case (project justification)
- Actively involved in the chartering process
- Assigning a project manager
- Assuring that the project team agrees on clear and measurable metrics
- Approving major project deliverables
- Participating in project governance and business-issue resolution
- Approving user training and documentation
- Motivating and inspiring the project team
- Overseeing risk-mitigation planning
- Recognizing and rewarding project success
- Agreeing on the alignment between project and organizational strategy
- Providing information on project impact, degree of change, organizational readiness, parallel/dependent projects, and constraints

(Source: Project Management Excellence published by Paton Press, page 27)

Project Team Leaders

Project team leaders should be accountable to the project manager. For each group of special skills on the project team, there is usually a project team leader. Although project team leaders have specific skills in a particular area of business or technology, they are not the same as subject-matter experts, as they accomplish specific objectives through the successful application of skills from other workers.

Project team leaders will also usually assist with the coordination and scheduling of activities, preparing detailed project work plans, actively participating in problem solving and conflict resolution, and identifying the resources best suited to performing the required project tasks. Project team leaders will also act as the primary liaison between the project manager, other project team leaders, and their assigned team members.

Team Leaders

Team leaders are different from project managers, as there is only one project manager on a project, but there can be many team leaders who are responsible for different teams that work on the same project. The team leader should be accountable to the project manager as are all project team members. For each group of special skills on the project team, there is usually a project team leader.

Team leader responsibilities usually include the following activities:

- Select team members and identify their roles and responsibilities
- Recommend meeting times and agendas
- Oversee work assignments
- Identify discrete team deliverables, accountability, timelines, and integration
- Provide reliable and timely performance status to the project manager
- Be encouraging, supportive, and tolerant of mistakes
- Keep the team focused on tasks at hand
- Take the steps necessary to keep team progress moving forward
- Surface issues to the project manager
- Be a good listener and motivate team members to do their best

Project Managers

A project manager is an organizational employee, representative, or consultant appointed to prepare project plans and to organize the resources required to complete a project, prior to, during, and upon closure of the project lifecycle.

- Definition of project manager: The project manager is the individual responsible for managing a project – responsible for the overall project and its deliverables; acts as the customer's single point of contact for services delivered within the scope of the project; and controls planning and execution of the project's scope for activities and resources toward meeting established cost, time, and quality goals.

It is said that a project manager is a person who accomplishes unique outcomes over critical timelines with limited resources, in order to meet the organization's objectives. I liken project management to quality management. Everyone thinks they know what quality is, so anyone can manage quality; this same thought pattern applies to project management. But just as a quality manager is a special type of professional with very special skills and training, so is a project manager. Project managers require skill, training, and effective leadership specifically related to project management.

Projects that are selected should have project managers immediately assigned for accountability. This individual usually reports either directly or indirectly to the project sponsor. A project manager's primary role is to manage the day-to-day project activities. This includes leading, directing, and coordinating activities related to planning, executing, and implementing projects. Many of these activities have to be managed simultaneously: scope, schedules, plans, budgets, process trends, project metrics, change requests, team dynamics, stakeholder relations, organizational change, and the facilitation skills associated with root cause analysis and corrective action as required. Therefore, it is vitally important that the project manager has senior management support, and that assistance should mean total buy-in and trust. Only through senior management support will the project manager have the authority to perform his or her duties in a situation of accountability (for the project) without having direct authority over the resources that are performing the activity.

Some of the day-to-day activities of the project manager include, but are not limited to, the following:

- Coordinating approval of project changes, risks, and issues
- Improving collaboration between teams throughout the project lifecycle
- Implementing a consistent and repeatable approach for project delivery

- Leveraging scarce resources to best utilize valued skills and experience
- Attaining commitment from project team members to their assigned deliverables
- Providing reliable and timely distribution of information to project stakeholders
- Directing the project to completion in an orderly and progressive manner
- Ensuring that the trade-offs between performance, scope, and costs are satisfactory
- Maintaining oversight of project performance, schedule, cost, and staffing
- Delivering the proper amount of coaching and mentoring techniques to develop personal and professional skills of all project team members
- Following the project sponsor's directions

Organizations are looking to their project managers to help meet the incredible challenge of remaining competitive in this global marketplace. The rapidness of change has created the need for large numbers of projects to be launched simultaneously. So, in many cases, project managers have more than one project to manage, causing resource constraints across the entire organization. To prepare the project manager for the many challenges he or she must face during the project lifecycle, senior management must provide the training, practical methodology, tools, leadership strategy, and empowerment for the project manager to make decisions wisely and effectively.

As "agents of change," project managers are also expected to transition an organization to a future state of operation while reducing cultural resistance. That type of success requires cross-enterprise cooperation. Few people are born with that leadership skill, so most project managers have to be properly trained in the areas of resistance, consulting, and overcoming obstacles. In other words, project management is not just about work plans and milestones – it's mainly about leadership.

Some of the more important leadership attributes for project managers are as follows:

- Listening skills
- Charisma

- Ability to motivate people
- Ability to get along with senior management
- Problem identification and solution
- Flexibility and adaptability
- Conflict resolution and negotiation skills
- Creativity
- Mentoring and coaching skills
- Financial ability
- Management ability
- Consulting skills

How can your project managers ensure that their projects will be delivered on time and on budget if we acknowledge that less than 26% of IT projects are successfully completed on time and on budget; 46% are late, over budget, or fail to meet the defined scope and quality; and 28% are cancelled? Here are some suggestions:

- Capture and share project intellectual capital:
  - Access, update, and store "post-project process review" reports, standard templates, sample projects, and best-demonstrated practices
  - Respect formalized project management methodology/flow
- Establish integrated project documentation management:
  - Control revision and change
  - Audit trail project documents
  - Author, modify, store and retrieve templates
  - Monitor concurrent access and modifications
- Manage work breakdown structures (both top-down and bottom-up):
  - Track projects through a single database repository across multiple functions within the organization on a global basis
  - Request, negotiate, and accept resources scheduling in cooperation with the resource manager using generic profiles
  - Roll up all budgets, incurred costs, work effort, resource allocation; and schedule to project level, initiative level, and portfolio level
  - Define links at multiple levels to reflect resource constraints or dependencies of deliverables and outputs

- Identify and perform strategic trade-offs:
    - Scenario analysis across multiple projects to forecast results on timing, resource allocation, and project status/progress
    - Review project's critical path, identify project milestones, and deliverables
    - Report on performance against baseline at all levels
- Quantify, qualify, and respond to changes, risks, and issues (CRI):
    - Track CRIs through identification; assignment of responsibility, dates, classification, priority level and response tasks; and impact evaluation and resolution
    - Store CRIs in the central database and retrieve them at multiple levels
- Ensure projects and portfolios are performing within budget:
    - Estimate budgets, building on readily available historical data
    - Capture internal and external staff effort, procurement, fund, and non-personnel costs
    - Allocate and charge-back expenditures across business units, functional groups and cost centers, projects, and portfolios
    - Plan, control, and monitor incurred and committed costs, payments, revenues, purchase orders, allocation of materials, and shipping/receiving of project materials
    - Allow for fixed, unit costing and amortization of capital expenditures
- Report to management on project development:
    - Schedule predefined, automatically broadcasted reports
    - Generate an array of performance, project status, and executive summary reports at multiple levels

The following is a typical project manager's weekly checklist:

- Validate project actuals
- Update project workplans with actuals
- Update budget with actuals
- Update control plan with actuals
- Review issue log and address outstanding issues
- Review Change Request log
- Update project statistics/performance data

- Determine status
- Prepare weekly project status report
- Forecast project completion based on status
- Initiate action to keep/bring projects back to plan
- Update short-interval schedule in the work plan
- Hold team meetings
- Meet with a project sponsor

The most valuable and least-used phrase in a project manager's vocabulary is "I don't know."

Innovation Project Team

The team members are the lifeblood of the team. The idea of "Participative Management" is based on allowing employees to help management make better decisions. The whole concept of "synergy" is based on two heads being better than one (1 + 1 = 3). But be careful. I have seen teams where 1 + 2 = 0.3 because the best ideas are compromised so everyone will agree. If the team leader is there to guide, the team members should assume responsibility for successfully completing the task.

Some specific team member responsibilities are:

- Willingness to express opinions or feelings
- Active participation
- Listening attentively
- Thinking creatively
- Avoiding disruptive communication
- Willingness to call a time-out when necessary
- Being protective of the rights of other members
- Being responsible for meeting the goals and objectives of the team

In a Six Sigma system, team members are typically Yellow or Green Belts. At times people who have not even been trained to the Yellow Belt level will be members of a Six Sigma team because they have specific skills or knowledge. Sometimes Black Belts will serve as a team member of a team that is led by another Black Belt or Master Black Belt.

TEAM GROUND RULES

My personal number one rule is "No one speaks more than they listen. If you speak for five minutes, you had better listen for five minutes or more but never less."

—H. James Harrington

The ground rules are a set of standards of behavior and attitudes that the team agrees to abide by. They should be established in the beginning stages of the team's formation and should include expectations regarding:

- Punctuality
- Respect for team members
- Member responsibilities/commitments
- Meeting etiquette
- Ability to listen
- Juggling team tasks with normal work tasks

The entire team is responsible for seeing to it that the ground rules are established and followed. Figure 4.1 lists the typical things that should be addressed in the ground rules.

Figure 4.2 is a list of typical ground rules for all team members, including the team leader and the facilitator.

The team is the heart of the problem solution cycle. Without an excellent team, the Six Sigma system will fail. Bringing the team together and making it hum is the most important part of the total system. Green Belts too often focus on technical tools and skip over the softer side of team building. Don't let that happen to you. Remember that when you are on a team, you're responsible to help the team function. Do your part even if you aren't the team leader (Figure 4.3).

Ground Rules

Attendance

- Arrive on time, stay until the end.
- Start on time, finish on time, unless otherwise agreed.
- Make every effort to attend. Provide advance notice to the meeting chairperson if you cannot, and arrange for your representative to attend in your place.
- No unannounced interruptions.

Objectives and Agenda

- Publish an agenda and stick to it unless otherwise agreed.
- Clearly state the objective for the meeting and stay focused.
- Use the group's time wisely; deal with the most important issues, or share information of general interest.

Communication

- Use active listening: Recognize that every idea or concern may be valuable. Do not dominate the conversation.
- Be concise.
- Use sensitivity: Question, do not challenge.
- Participation is essential. Express your honest viewpoint or concern. No hidden agendas.
- Focus *constructive* criticism on exposing or removing obstacles.

Respect and Courtesy

- Respect and understand others' positions and feelings.
  - Avoid behaviors, which discourage involvement:
  - Interrupting
  - Side conversations
  - "Killer" phrases, body language, or gestures
  - Not paying attention

FIGURE 4.1
Typical Ground Rules for Team Leaders (Continued)

Teamwork

- Explore and disagree within the meeting. Present a single approach outside.
- Strive for consensus. Use appropriate decision-making tools if necessary. Resolve irreconcilable differences outside the meeting.
- Have some fun: Some teams set up $1.00 fines for improper behavior. They use the money to buy pizza for the team meetings.

Support and Follow-up

- Review meeting effectiveness periodically.
- Address problems openly.
- Assign a facilitator to monitor team processes and behaviors.
- Assign a scribe to record and issue minutes.
- Assign a timekeeper to keep the meeting on schedule.
- Make clear action assignments and carry them out.
- Use visual media (flipcharts, overheads) to share ideas.

FIGURE 4.1 (CONTINUED)
Typical Ground Rules for Team Leaders

	Session Behavior Standards ***Session Leader***
1	Start breaks on time (± 10 minutes maximum)
2	Start session on time (-5 minutes maximum)
3	Follow the agenda
4	End session on time (± 15 minutes maximum)
5	Don't interrupt
6	Keep accurate error-rate logs
7	Send agenda out 3 days in advance
8	Document meeting within 48 hours

FIGURE 4.2
Session Behavior Standards for Session Leader

	Session Behavior Standards for Attendees
1	Arrive on time (-5 minutes maximum)
2	Back from breaks (-2 minutes maximum)
3	Back from lunch (-2 minutes maximum)
4	Ask 1 question or comment at each meeting
5	Come to the meeting with your assignment completed and with copies of the results for the other members
6	Encourage others to express their opinion
7	Never belittle another team member's ideas. Build on them to make them better if you can
8	Document meeting within 48 hours
9.	Stay to the end of the session
10.	Don't interrupt
11.	Don't monopolize the discussion
12	Keep accurate error-rate logs

FIGURE 4.3
Typical Ground Rules for the SS Team

5

The PIC Overview

ISO 56002:2019 INNOVATION PROCESSES CYCLE

ISO 56002:2019 suggests that the organization's innovation processes be modified to suit the individual initiative, depending upon the type of innovation and the circumstances within the organization. To reflect this thought pattern in Subclause 8.3.1 General, they show their concept of an overview of the innovation processes (see Figure 5.1).

The five ovals in Figure 5.1 represent five subclauses of ISO 56002:2019:

- Subclause 8.3.2 Identify opportunities
- Subclause 8.3.3 Create concepts
- Subclause 8.3.4 Validate concepts
- Subclause 8.3.5 Develop solutions
- Subclause 8.3.6 Deploy solutions

> You will note that in ISO's view of the innovation processes, the intent goes directly to identifying opportunities. From identifying opportunity, the activity flow goes directly to create concepts. From create concepts, the activity flow goes directly to validate concepts. Although there are three outputs from validate concepts, the most logical path is to go directly to develop solutions and then to deploy solutions. The output from the ISO innovation processes is value. ISO defines a value can be either or both financial and nonfinancial. (©ISO. This material is excerpted from ISO 56000:2020, with permission of the American National Standards Institute (ANSI) on behalf of the International Organization for Standardization. All rights reserved.)

To explain the complexity of the innovative processes, there are two alternate outputs from validate concepts. One goes back to identify opportunities, and the other goes directly to deploy solutions. There is

DOI: 10.4324/b22993-5

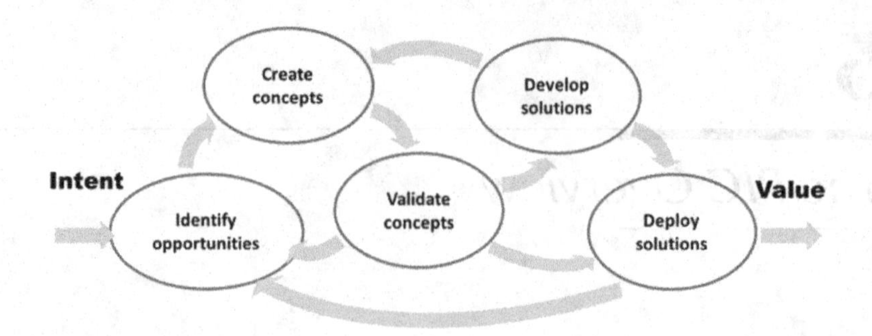

FIGURE 5.1
ISO 56002:2019 Overview of the Innovation Processes (©ISO. This material is excerpted from ISO 56000:2020, with permission of the American National Standards Institute (ANSI) on behalf of the International Organization for Standardization. All rights reserved.)

only one alternate path for the output from develop solutions and that is back to create concept. Deploy solutions also have one alternate output path and that is back to identify opportunities.

The *Enhance* cloud service at **EDGESoftware.cloud/managing-innovation** supports just such customization. You can start with the Best Practice processes being presented in this book and modify them to adapt to your particular needs for each project. Then you can manage the project as it progresses through your custom processes.

PIC PHASES

There is no wrong or right way to divide up the IMS into a sequence of sub-processes/activities. The IMS is a very complicated grouping of many processes, tasks, and activities together that frequently interact with each other throughout the cycle.

Our view of the innovation process cycle is slightly different from ISO 56002:2019 subclause approach. ISO's approach provides a system view of the cycle; we have decided to focus on a project view of the cycle. This has resulted in a three-phase approach that is made up of 12 Process Grouping and five Tollgates. This approach focuses on what it takes to use the system and maximize the results from using the system. As the standard name implies, IMS focus has been on developing the system. The approach that we are using is on the innovation project cycle. This provides a focus on how you maximize the positive results and decrease the number of failures related to your innovation projects.

In this book, we will be using the three-phase approach divided into 12 Process Groupings approach (listed below) because it has more detail focused on the latter part of the innovation project cycle, and the Process Groupings provide additional focus on key operational points. Our surveys indicate that the truly successful innovative companies use a wide range of systems and put prime focus on individual innovative initiatives to accomplish their organizational goals.

- Process Grouping definition – Process Groupings are groups of processes and systems that are gathered together due to the way they interact with each other or, in some cases, are alternative processes. (For example, you can do something by hand or electronically. In either case, the activities involved are very different.)

In the case of the PIC, it is made up of 12 Process Groupings (see Figure 5.2). These 12 Process Groupings can be viewed in the *Enhance TeamPortal* (see Figure 5.3).

Project Innovation Cycle (PIC)

12 Innovation Process Groupings

Phase I. Creation

- Process Grouping 1. Opportunity Identification
  *Tollgate I - Opportunity Analysis
- Process Grouping 2. Opportunity Development
- Process Grouping 3. Value Proposition
  *Tollgate II - Concept Approval
- Process Grouping 4. Concept Validation

Phase II. Preparation and Producing

  *Tollgate III - Project Approval
- Process Grouping 5. Business Case Analysis
- Process Grouping 6. Resource Management
- Process Grouping 7. Documentation
- Process Grouping 8. Production
  *Tollgate IV - Customer Ship Approval

Phase III. Delivery

- Process Grouping 9. Marketing, Sales, and Delivery
- Process Grouping 10. After-Sales Services
- Process Grouping 11. Performance Analysis
  *Tollgate V - Project Evaluation
- Process Grouping 12. Transformation

FIGURE 5.2
PIC Process Groupings and Tollgates

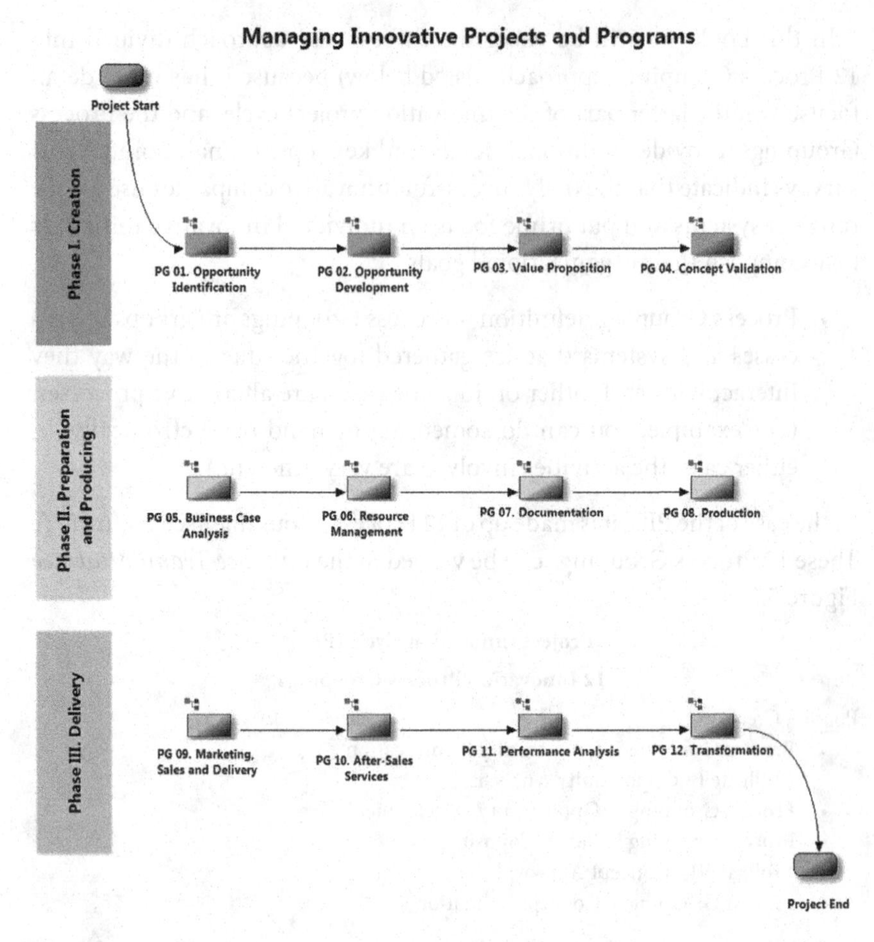

FIGURE 5.3
PIC Process Groupings as seen in EDGE Software's *Enhance TeamPortal* cloud service

Each of the 12 Process Groupings will be discussed in detail in Chapters 6, 7, and 8. In addition, activity block diagrams have been prepared for each phase that will help you consider activities that will be beneficial for the project you are involved in. We selected using activity block diagrams rather than flowcharts because activity block diagrams allow bigger sections of the system to be pictorially combined, and the narrative descriptions related to the individual activity provides a more detailed understanding of the system.

PIC TOLLGATES

Due to innovative projects' high failure rates (estimated to be as high as 75%), we have built into the cycle five review points called Tollgates. We firmly believe that the right time to stop a project that is going to fail is as early in the cycle as possible – even better, before the cycle begins. We suggest that you use the Tollgates as celebration points as well as go/no-go decision points. The basic focus of the Tollgate agenda is the recognition of the accomplishments the team has made and the potential obstacles the team will have if the project continues. The real failure occurs when a project gets completed and outputs are being delivered to a customer/consumer before the organization is sorry that they started the project.

There is a basic thought pattern related to innovation projects where failure is considered a learning experience rather than a negative incident. I personally have a difficult time accepting this. I admit that if the project fails, you should look for what good things came out of it and what you gained out of it. Probably the most obvious is the ability to learn from the mistakes we make. All too often I learned from my mistakes, but learning when you fail is knowledge only if it prevents it from happening again. Communicating learning with the project fails as this knowledge is now in the product of the effort. I think you will admit that learning from our failures is an awfully expensive way to get educated, and it should not be considered as an employee right. It's like saying, "Well I didn't marry the prettiest girl in the class, but I married the one with the smallest feet." Or "I didn't marry the smartest man in the class, but I married the one from the biggest family."

- Definition of Tollgate: Tollgate in project management is a virtual gate between the project phases. It closes one set of activities – if and when it is classified as complete – and opens the next set of activities of the project. It is important to understand that Tollgate is usually a moment in time when a business decision is made.

We had a great deal of discussion about the value and negative impacts Tollgates have on individual projects. Some of my clients feel that it's a complete waste of time and detracts from the get-it-done activities, while others feel equally as strong that it's the thing that gets things done on schedule and within costs. The truth of the matter is both of them are right and both of them are wrong. In cases like, "What should you do?" I like to make a list of good and bad points and then make a decision.

Bad point perspectives: Why shouldn't we have Tollgates?

1. It adds additional unneeded work to the project.
2. It wastes a lot of executive time.
3. It slows the project down.
4. It adds additional unneeded stress to the project.
5. It requires additional record-keeping.
6. It's a big waste of time on small projects.

Good point perspectives: Why should we use Tollgates?

1. It keeps the executive team interested and knowledgeable about the projects.
2. Everyone gets a good understanding of where the organization is going.
3. It reduces cost overruns.
4. It helps keep the project on schedule.
5. It eliminates big surprises.
6. It demonstrates executive concern and interest.

Okay, you see what I mean. I'll bet you can add two or three more to each of these lists that reflect your thought patterns in your organization. I personally like Tollgates, and I feel they are beneficial if used correctly. I like using Tollgates for the following reasons:

1. It is the activity to get more meaningful and factual information.
2. It is the activity to recognize the progress my team has accomplished.
3. It is the activity to resolve a disagreement between team members.
4. It is the activity to improve communication throughout the organization.
5. It is the activity to terminate weak projects.
6. It is the activity to help prioritize my work.
7. It is the activity to help develop some of my high potential employees.
8. It is the activity to help with the change management activities.
9. It is the activity to meet and see individuals perform that I would not normally see in my management-by-walking-around approach.

I have frequently thought that we look at Tollgates as a negative activity because we associated it with a Tollgate where we pay money to allow us to drive over a bridge. I personally believe it should be a celebration mark where the individual in the project team has a chance to brag about the fine work their team has been doing. For items that are not on schedule, the focus should not be on why they aren't schedule, but what does the group need to do to bring it back on schedule. We seriously considered dropping the term "Tollgate" for "Project Accomplishment Reviews (PARs)." These reviews should focus on what we've accomplished and address what we can do to get the ones that are not on schedule back into tune with the rest of the project. We decided not to make the change because it would just add additional confusion to the already-complex IMS.

I used the following five rules in establishing the Tollgates:

1. I funded projects only until the next Tollgate.
2. The Tollgate is directed at things that need to be accomplished at a specific point in the project.
3. The total Tollgate approach is only used on major projects.
4. Projects that do not warrant the use of project management or require executive-level surveillance will have fewer Tollgates.
5. Some of the five Tollgates are presentations to the executive sponsor of the project only.
    - Tollgate I. Opportunity Analysis
    - Tollgate II. Concept Approval
    - Tollgate III. Project Approval
    - Tollgate IV. Customer Ship Approval
    - Tollgate V. Project Evaluation

It is apparent that project evaluation should take place. Logically it should be conducted after Process Grouping 11. Performance Analysis report has been completed, as it will serve as one of the key inputs to this Tollgate. The purpose of this evaluation is to determine if the improvement opportunity had been properly addressed and if the commitments made in the mission statement and project plan have been fulfilled. This evaluation should go beyond the entities' performance and include the estimated versus actual resources used and cycle time.

At this point in time, the analysis should focus on how the change impacted each of the types of stakeholders. For example, "What was the change done to customer satisfaction? Did it impact safety? How was corporate culture impacted? How are the surplus resources handled? What were the lessons learned? How was it communicated to the rest of the organization? And how did it impact markets?"

Typically, the operating procedure for Tollgate V would include a checklist of the supporting potential impact areas that need to be addressed in the Tollgate V presentation and documentation.

Often there is a great deal of reluctance to conduct Tollgate V and even Process Grouping 11. Performance Analysis. I can't tell you how many times I've heard CEOs saying, "That project is over. Let's get on doing something useful." Certainly, there is a need to free up these key knowledgeable employees and get them involved in a different opportunity. Because of this, it is sometimes effective to reassign all but one or two members of the team, leaving them to collect the measurement data, provide key input into the knowledge base, prepare the final report, and have the executive sponsor host the Tollgate V meeting. I believe it is important to bring the whole project team back together for the Tollgate V meeting if possible. This is the single point in time that management can thank project teams for their accomplishments and their ability to work together to move the organization forward. Remember this is a celebration of an accomplishment, not a funeral or witch-hunt.

Five Tollgates Summary

Be careful not to look at the Tollgates as a negative exercise. Quite the contrary; this should be treated as an opportunity to recognize the progress and creativity of the project team. The key to success for any team in business or games is the thrill of winning. As we approach the PIC, we should look for the three drivers that measure personal success. It's the same in all games. These three drivers are:

1. You need to have a set of rules that everyone is using.
2. You need to have a way of measuring your progress. Just think how boring it would be if you were just bouncing the tennis ball against the garage door versus a normal tennis match.
3. You need to have a goal that's worthwhile striving for. I bowled for years with the goal of having a 300 game and never quite made it. But although it was 60 years ago, I still remember the game I bowled 260.

The Tollgates provide this point of recognition that we all want at the end of a hard-fought game; we did our very best. It's the point in the cycle when management should be recognizing the team's accomplishments rather than focusing on the team's failures. I make it a rule at these Tollgates meetings to use the word "how" rather than "why." Example: Don't ask, "*Why* are you behind schedule?" The right question is, "*How* can we get back on schedule?"

INTRODUCTION TO PIC THREE PHASES

Now that we have defined the cast of players and the role the major players will play, we are ready to start explaining the three phases that make up the PIC.

- Phase I – Creation
- Phase II – Preparation and Producing
- Phase III – Delivery

We will devote the next three chapters to each of the three phases. Each phase will be treated as comprising unique entities that can interact with any one of the other 11 Process Groupings. For example, Resources is part of Phase II, but all three phases need to have a continuous flow of resources in order to complete their assignments.

It may be easier to understand if we use a concrete example, so we developed activity block diagrams for the PIC of an electromechanical device for all three phases. It will include a mid-level analysis as most of the major activities are conducted by the functions involved in the project. We will be working with basic assumptions that may or may not be in compliance with the ISO International Standard 56002:2019 – Innovation Management System. We find that most companies already have installed an IMS that is in compliance with ISO's definition of innovation. Some of them have proven to be very effective; others are just limping along implementing enough corrective action to keep going. I frequently hear people say, "Americans are the best problem solvers." That's probably true because we have more practice.

It is also very helpful if you already have the following methodologies in place and functioning efficiently and effectively.

- Organizational change management
- Knowledge management
- Value-added analysis system
- Project management systems

If that is not the case, there can be a great deal of variation in success levels from project to project, and the project cost and the project cycle time will be much greater than they should be. It's like a 4 × 100 meters relay track race at the Olympics where one country hands off the baton to four different runners while another country tries running the total 400 meters with just one athlete. It can be done, but it is much more difficult, riskier, and the probability of success is much lower.

This chapter will describe the basic processes that make up the PIC. The complexity of the PIC makes it difficult and confusing to flowchart all of the processes in detail by hand.

In the next three chapters, we will be developing activity block diagrams for moving projects through an already-established IMS.

Up to this point in the book, we have focused on the requirements for innovation and Phase 1. Creation of the PIC. This is the phase that everyone likes to write about and talk about. It's like a courtship, marrying, and honeymoon of the man-woman relationship. It includes four Process Groupings.

- Process Grouping 1. Opportunity Identification
- Process Grouping 2. Opportunity Development
- Process Grouping 3. Value Proposition
- Process Grouping 4. Concept Validation

Completing these four Process Groupings should result in the individual or team feeling a sense of accomplishment for originating a potentially innovative opportunity. It's a lot like the feeling a fullback experiences on the football team when the player crashes the opposition's line breaking through to make a touchdown. The player receives a strong sense of accomplishment, self-satisfaction, and recognition by his or her team.

As we explained earlier, we are using the term "Process Groupings" because in each case, there are a number of processes that come together or exist in parallel to deliver the desired result. For example, there are many different processes involved in preparing the value proposition. The finance department has one or more groups of processes as does the marketing department. The same is true of functions like manufacturing engineering, manufacturing, quality assurance, sales, product engineering information technology, etc.

It seems that everybody hates documentation. We hate to take the time to write it down. We hate to take the time to convince somebody else that this is the right thing to do. We want to have a paperless organization, but we still need some way of communicating what should be done, when it should be done, and how it should be done. The basic information should be stored in a knowledge database which is used to provide direction and allows everyone to benefit from the good and bad experiences that the individuals within the organization have experienced.

As has already been mentioned, the *Enhance* Public Knowledgebase that serves as a "live extension" to this book already offers a tremendous amount of documentation in the form of the Best Practice processes you are learning about in this book. *Enhance* allows these processes to be customized for each project that you run, and as an added benefit you can add your own supporting knowledge assets in the form of documents, book extracts, links to web-based resources, and even videos to your customized processes. For example, in a manufacturing environment, these knowledge assets might include:

- Job specifications
- Lists of necessary tools
- Required equipment
- Required fixtures
- Measurement equipment
- Measurement procedures
- Training plan

SIX LEVELS OF DOCUMENTATION

In most organizations, there are five or six levels of documentation. Each level breaks the previous level down into more detail. There are two basic approaches to designing your documentation system levels (see Figures 5.4 and 5.5).

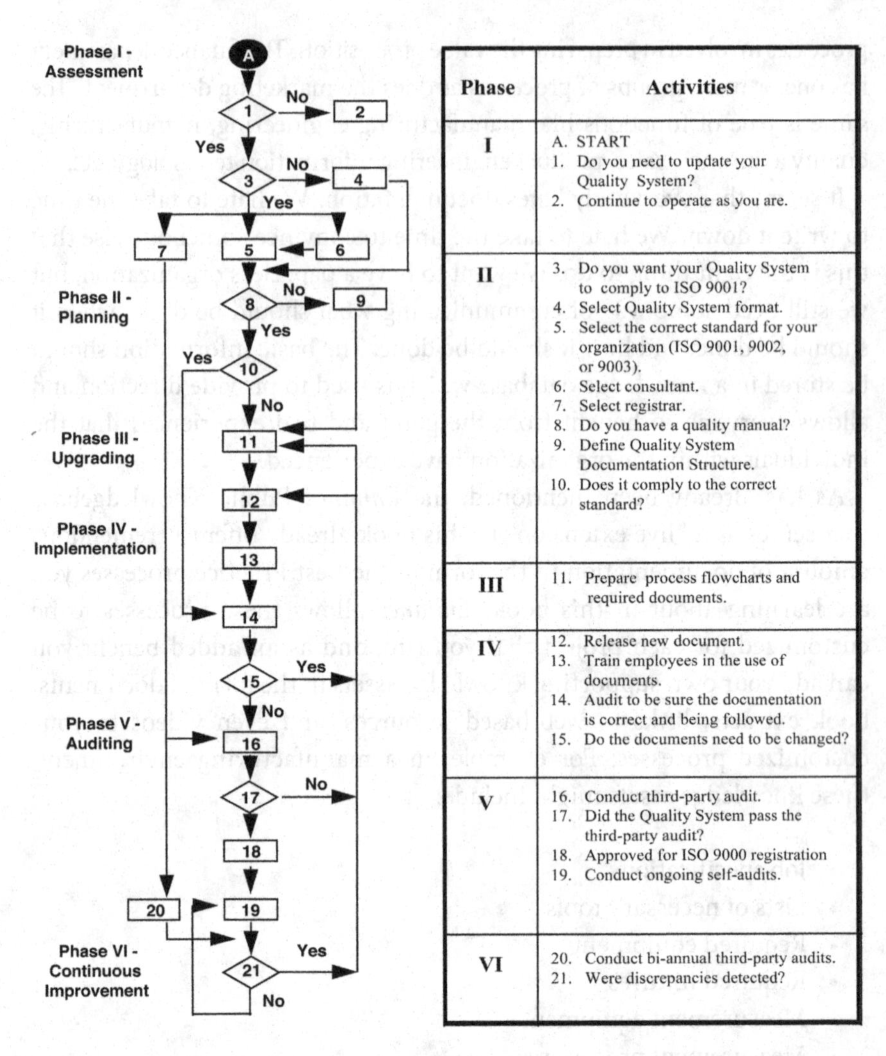

Phase	Activities
I	A. START 1. Do you need to update your Quality System? 2. Continue to operate as you are.
II	3. Do you want your Quality System to comply to ISO 9001? 4. Select Quality System format. 5. Select the correct standard for your organization (ISO 9001, 9002, or 9003). 6. Select consultant. 7. Select registrar. 8. Do you have a quality manual? 9. Define Quality System Documentation Structure. 10. Does it comply to the correct standard?
III	11. Prepare process flowcharts and required documents.
IV	12. Release new document. 13. Train employees in the use of documents. 14. Audit to be sure the documentation is correct and being followed. 15. Do the documents need to be changed?
V	16. Conduct third-party audit. 17. Did the Quality System pass the third-party audit? 18. Approved for ISO 9000 registration 19. Conduct ongoing self-audits.
VI	20. Conduct bi-annual third-party audits. 21. Were discrepancies detected?

FIGURE 5.4
Third-Level Activity Block Diagram for Installing ISO 9000

Figure 5.5 shows how the levels interact with the previous level in the departmental level design.

Both the process and the departmental approaches work equally well when applied correctly.

Organizations that are smokestack generally use the departmental approach.

Those organizations that are process-focused use the process-focused approach. We personally prefer the process-focused approach as it better defines and documents the way the organization is functioning. This

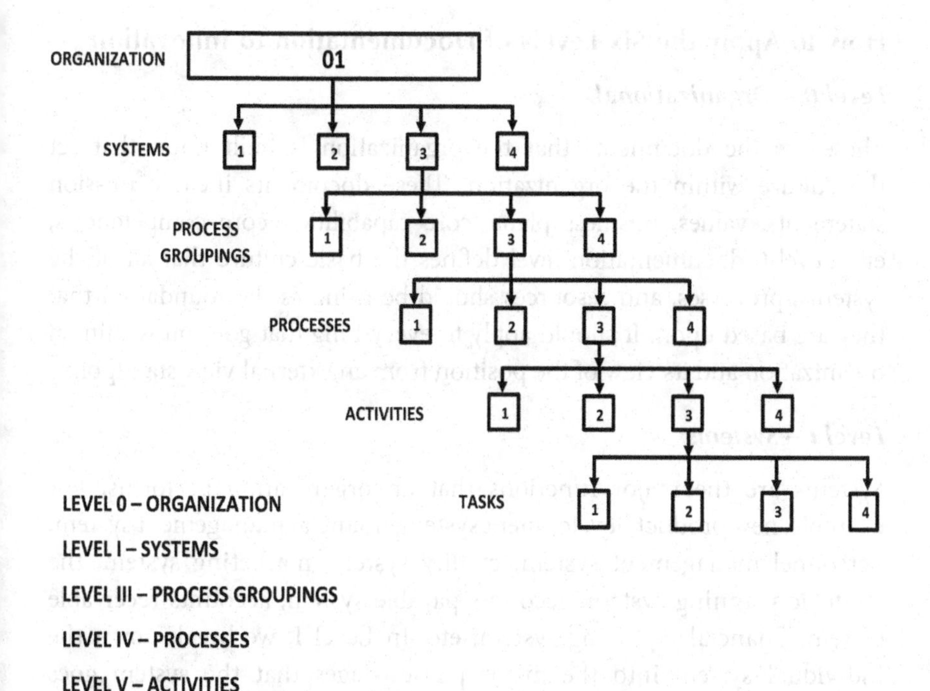

FIGURE 5.5
Process Documentation Levels Interaction

offers the advantage of focusing on how the process is functioning rather than how individual departments are functioning. Often focusing on individual departments can result in suboptimization of the total process performance. As a result, we will be using the process approach as an example throughout the remainder of the book.

At this level, the individual processes are broken down into activities that are recommended in order to have the process function properly. For example, processes are often used to define the movement of output within the organization; they can be used for products, information, records, etc. Flowcharting is a common method for laying out and documenting a process. In our case, we are using activity block flowcharting sometimes called activity block diagrams. The blocks within the block diagram are known as activities. In summary, at Level IV, each major process is divided into the activities that are necessary to successfully complete the process.

How to Apply the Six Levels of Documentation to Innovation

Level 0 – Organizational

These are the documents that the organization is built upon that set the culture within the organization. These documents include mission statements, values, business plans, core capabilities, core competencies, etc. Level 0 documentation level defines the basic culture that all of the systems, processes, and resources should be using as the foundation that they are based upon. It should apply to everything that goes on within an organization and its view of the position from an external view standpoint.

Level I – Systems

Systems are the major functions that an organization performs. For example, new product development system, financial management system, personnel management system, quality system, marketing system, the strategic planning system, accounts payable system, accounts receivable system, financial reporting system, etc. In Level I, we break down the individual system into the major phases/stages that the system goes through in the system's lifecycle. Most systems will at least go through a development phase, an evaluation phase, an implementation phase, an operational phase, and an upgrading phase. In summary, Level I lists the major systems that are used to operate an organization.

Your effort to improve an organization's innovation at the Level I – Systems could be called an "Innovation Management System." It could be subdivided into three phases as shown in Figure 5.6.

You will notice that the systems define a new output's lifecycle. This often starts with recognizing the opportunity, preparing the organization to make the output, following through with delivering the output to the internal or external customers, and ending with the customer support activities. This basic model should serve as the starting point for designing all the organization's systems and processes whether it services just internal, external, or both customers.

System #1– Innovation Management System (Level I)

- 1.1 Phase I – Creation (Level II)
- 1.2 Phase II – Preparation and Production (Level II)
- 1.3 Phase III – Delivery (Level II)

FIGURE 5.6
Innovation Management System Divided into Three Phases

Level II – Phase Levels Are Divided into Process Groupings

Each of the phases is divided into Process Groupings that define the way that the phase can be completed successfully on time and at minimum cost. Each individual Process Grouping can contain as little as one process or more than 100 processes for really complex phases. In summary, Level II breaks the individual processes down into lifecycle time periods called Process Groupings. In some cases, for Level II, department names might be used instead of Process Groupings.

At Level II, each of the phases is divided into the Process Groupings and Tollgates that are required to start, perform, and complete the phase. In some cases, a Process Grouping may be made up of only one process, but that is not normally the case. In most complex systems, a number of individual systems will need to work together to provide the results that are required from each phase of the output's lifecycle. The following shows how Phase II – Preparation and Producing can be divided into four Process Groupings and two Tollgates (see Figure 5.7).

Tollgate IV Customer Ship Approval is shown to occur after Process Grouping 8. Production because when Tollgate IV is complete, all entities that are produced should be in compliance with the specifications and customer requirements, and can be directly shipped to external customers. Output produced before the successful completion of Tollgate IV should not be shipped to external customers; that is, unless the customer is told in advance that this is reproduction produce items and may or may not perform as specified or to their satisfaction. This Tollgate should be the very last activity in Phase II – Preparation and Producing.

In this case, Phase II – Preparation and Producing is divided into six Process Groupings. Each of these Process Groupings will contain one or more processes that are required to complete the Process Grouping objectives. You should note there are two Tollgates included in Phase

1.2. Phase II – Preparation and Producing (Level II)
- 1.2.1 Process Grouping 5 - Business Case Analysis (Level III)
- 1.2.2 Tollgate III - Project Approval (Level III)
- 1.2.3 Process Grouping 6 - Resource Management (Level III)
- 1.2.4 Process Grouping 7 - Documentation (Level III)
- 1.2.5 Process Grouping 8 - Production (Level III)
- 1.2.6 Tollgate IV - Customer Ship Approval (Level III)

FIGURE 5.7
Process Groupings and Tollgates for Phase II. Preparation and Producing

II: the first Tollgate is where the project gets approved as being part of the organization's project portfolio; and the second one verifies that the output will be in keeping with the documentation, all of the objectives in Phase II have been successfully completed, and the product is ready to be delivered to its internal or external customer. In cases where there are few processes involved in the system, Level II can be skipped and the process can be linked directly to the system.

Level III – Process Groupings Are Divided into Processes

At this level, the individual Process Groupings are broken down into individual processes. For example, in the case of 1.2.3 Process Grouping 6 – Resource Management, all the major processes are identified that are needed to bring together the resources required to successfully complete the project/program objectives. This would include processes related to funding the computer systems, personnel required to operate the system, and equipment required to set up the system and to provide adequate space for the systems operations. (See Figure 5.8 for typical processes that would be included in 1.2.3 Process Grouping 6 – Resource Management.)

In this case, 1.2.3 Process Grouping 6 – Resource Management is divided into nine major processes that are required to bring together the resources needed to successfully complete the Process Grouping's objectives. This would include processes related to funding the system, personnel required to operate the system, equipment required to set up the system, and providing adequate space for the system's operations.

1.2.3. Process Grouping 6 – Resource Management (Level III)
- 1.2.3.1 Internal personnel availability process (Level IV)
- 1.2.3.2 Internal personnel skills analysis process (Level IV)
- 1.2.3.3 Hiring process (Level IV)
- 1.2.3.4 Budgeting process (Level IV)
- 1.2.3.5 Financial acquisition process (Level IV)
- 1.2.3.6 Procurement process (Level IV)
- 1.2.3.7 Systems space layout and installation process (Level IV)
- 1.2.3.8 Supplier selection/qualification process (Levels IV)
- 1.2.3.9 Subcontracting process (Level IV)

FIGURE 5.8
1.2.3. Process Grouping 6 – Resource Management Divided into Nine Major Processes

Level IV – Process Levels Are Divided into Activities

At this level, the individual processes are broken down into activities that are recommended in order to have the process function properly. For example, processes are often used to define the movement of output within the organization; they can be used for products, information, records, etc. Flowcharting is a common method for laying out and documenting a process. In our case, we are using Activity Block Diagrams that are a form of flowcharting. The blocks within the Activity Block Diagram are the activities that are conducted within the illustrated process. In summary, at Level V, each major process is divided into the activities that are necessary to perform the tasks.

Level IV takes each individual process and divides it up into activities. Because we want to view the total system, we need to tie together at least the major activities. For example, in Figure 5.8, individual major activities in the previous Level IV example entitled "1.2.3.3 Hiring process (Level IV)" could be made up of the activities listed in Figure 5.9.

1.2.3.3. Hiring Process (Level IV)

1.2.3.3.1 Workload analysis identifies additional human resources needed (Level V).

1.2.3.3.2 Obtain budget increase approval (Level V).

1.2.3.3.3 Search to determine if internal qualified staff is available (Level V).

1.2.3.3.4 Yes. Select best candidate

1.2.3.3.5 If no qualified internal resources are available, process external hiring requests (Level V).

1.2.3.3.6 Search for external candidates (Level V).

1.2.3.3.7 Personnel screen applicants (Level V).

1.2.3.3.8 Selects hiring candidates and schedule interviews (Level V).

1.2.3.3.9 Select best candidate & submit job offer.

1.2.3.3.10 Candidate accept assignment offer.

1.2.3.3.11 Complete contract with the new hire including health tests (Level V).

1.2.3.3.12 New employee available for training (Level V).

FIGURE 5.9
1.2.3.3. (Level IV) Hiring Process Divided into Major Work Activities

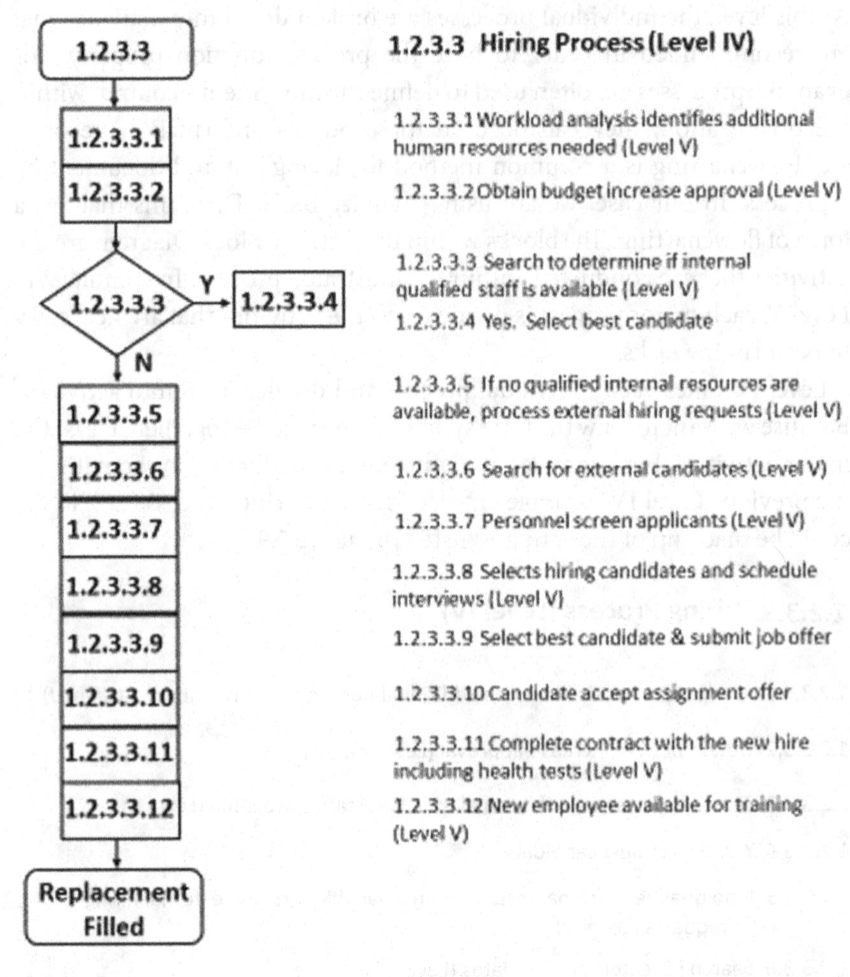
Activity Block Diagram (ABD) for Hiring Process (Level IV)
1.2.3.3
1.2.3.3.1
1.2.3.3.2
1.2.3.3.3
Y
1.2.3.3.4
N
1.2.3.3.5
1.2.3.3.6
1.2.3.3.7
1.2.3.3.8
1.2.3.3.9
1.2.3.3.10
1.2.3.3.11
1.2.3.3.12
Replacement Filled
1.2.3.3 Hiring Process (Level IV)
1.2.3.3.1 Workload analysis identifies additional human resources needed (Level V)
1.2.3.3.2 Obtain budget increase approval (Level V)
1.2.3.3.3 Search to determine if internal qualified staff is available (Level V)
1.2.3.3.4 Yes. Select best candidate
1.2.3.3.5 If no qualified internal resources are available, process external hiring requests (Level V)
1.2.3.3.6 Search for external candidates (Level V)
1.2.3.3.7 Personnel screen applicants (Level V)
1.2.3.3.8 Selects hiring candidates and schedule interviews (Level V)
1.2.3.3.9 Select best candidate & submit job offer
1.2.3.3.10 Candidate accept assignment offer
1.2.3.3.11 Complete contract with the new hire including health tests (Level V)
1.2.3.3.12 New employee available for training (Level V)

FIGURE 5.10
Hiring Process Flowchart

To accomplish grouping together all the activities from different processes, we are using a modified flow diagram technique called Activity Block Diagram and preferably with the activity description printed on the same document. (Note: In many cases, it is not practical to have the activity description printed on the same sheet of paper that the Activity Block Diagrams is recorded on.) See Figure 5.10 for an example of the hiring process in the Activity Block Diagram.

The items listed at the activity level V are often used as a heading in the project's work breakdown structure.

Level V – Activities Are Divided into Tasks

Level V takes the individual major activities and breaks them down into the tasks required to successfully complete the activity. It is the real details related to what a person needs to know or do. For example, it would define a tooling number that would be used, the specification that would be used, the equipment that would be used, the speeds and setting of the equipment, inspection procedure, and the videos showing how to do a specific job. In this book, we will not be going down to Level V.

At this level, the individual activities are now analyzed to determine what tasks need to be performed with what equipment in order to complete the activity and produce the activity's output. The tasks related to completing an activity include things like what equipment is used, what fixtures are used, how long will it take, what's required to do it, sequence of tasks, and prints specifications. In summary, Level V divides the individual activity into tasks (how-to directions) that need to be performed to successfully complete the activity.

By now, it should be obvious to you that a fourth-level and even fifth-level breakdown is required in order to include all of the work instruction information and direction required to perform a complex assignment. For example, if the individual was assigned to drill a hole in a casting, he should be provided with the correct size of the hole, tolerance, and location. He also would need to know which equipment he should use and how the equipment should be set up. With this additional information, he should be ready to complete his assigned task of drilling a hole in a casting.

To get to the task level (Level V), you take the individual major activities and break them down into the tasks required to successfully complete the activity. Tasks are the real details related to what a person needs to know or do the assignment. In this book, we will not be going down to Level VI. This information is normally stored in the computer program and made available to individuals when they're doing a specific activity. The computer program is designed to show all five levels to allow an individual to drive down from Level I to Level VI. Even Level V or Level VI information may not be required if a person is trained well enough to know what to do on an assignment without the documentation.

ACTIVITY BLOCK DIAGRAMS

The Activity Block Diagrams are simplified flowcharts that are designed to handle large quantities of information in a very systematic system-flow thought pattern. They use four basic symbols, although additional symbols can be added based upon customization needs. The four basic symbols are:

1. A rectangle with a number inside the rectangle that relates to the description in the activity list.
2. A diamond with a number inside the diamond that relates to the decision statement in the activity list. The decision outputs from the diamond decision are labeled on the direction line using one of these indicators:
    - Y = yes
    - N = no
    - P = past
    - F = failed
    - CA = corrective action
    - H = on hold
    - D = delayed
3. A rectangle with rounded corners indicates that it is either an input or an output. Usually, a statement where the output is going is included in the rectangle. On occasion, a simple number with the description included on the activity list is used.
4. Lines with an arrow at one or both ends of the arrow indicate the direction of activity flow from one rectangle or diamond to another.
5. They can be either vertical or horizontal Activity Block Diagrams.
    - The horizontal Activity Block Diagram makes the basic assumption that all flow is from the left-hand side to the right-hand side of the paper. In these cases, there is no need for a line or an arrow between the rectangle as long as the flow is directly

from left to right. Flow up or down does require a line and an arrow that indicates workflow direction.

- The vertical Activity Block Diagrams make the basic assumption that all flow is from the top to the bottom of the page. In these cases, there is no need for a line or arrow between the rectangles as long as the flow is downward. Activity flow that is upward or sideways does require a line and arrow to indicate work direction.

6

Phase I: Creation

INTRODUCTION

Phase I. Creation has the highest risk of the three phases. No matter what anyone says, we all know that every individual has a real sense of accomplishment when he or she comes up with a new and original concept that has a potential of adding value to some, if not all, of the stakeholders. After going through Phase I, the organization should have a potential innovation project.

There are four Process Groupings and two Tollgates that make up Phase I.

- Process Grouping 1. Opportunity Identification
- Process Grouping 2. Opportunity Development
- Tollgate I. Opportunity Analysis
- Process Grouping 3. Value Proposition
- Tollgate II. Concept Approval
- Process Grouping 4. Concept Validation

PROCESS GROUPING 1: OPPORTUNITY IDENTIFICATION

Identifying an opportunity that can bring about improvement within the organization and added value outside of the organization is a very positive experience. At this point, the individual thinks to himself or herself that this is an opportunity for me to do something that no one else has taken advantage of. It's my chance to be a white knight fighting to improve my performance.

DOI: 10.4324/b22993-6

> *Innovation Rule Number one – If it isn't broke, it's time to improve it.*
>
> **—H. James Harrington**

Phase I starts with Process Grouping 1. Opportunity Identification. This Process Grouping is responsible for identifying improvements and/or replacement opportunities. Most people have had the concept of strictly following the procedures and process without deviation implanted into their personality and habits resulting in becoming blinded to the potential of challenging the current status and breaking out from today's restrictions, rules, and regulations. They put blinders on, forcing them to look straight ahead down the process map without the ability to look left or right for improvement opportunities. Shattering this habit of strict compliance with established processes and procedures has greatly reduced the ability of today's generation to break away from the accepted document behaviors and be creative and innovative. The directives to strictly following established documented procedures as suggested by ISO 9000 have built a halo around our employees' heads, protecting them from any negative criticism and eliminating being responsible for failure.

The first step in instilling an innovative culture within an organization is training everyone on how to recognize improvement opportunities and why it is not only acceptable, but desirable, to challenge the way things are being done in order to find a way to do it better.

It's time to train all levels of management to welcome being challenged by their subordinates. Management needs to truly believe that being challenged is not a negative act but a positive one that they will thrive on.

It's time for management to accept that having a bad idea is not wrong; what is wrong is not having an idea at all. It is a good practice to hold a four- to eight-hour class on creative thinking. This should be a basic requirement for all employees and double that for management.

It's really not difficult to identify improvement opportunities. All you have to do is open your eyes and look around. Everything you look at, say to yourself, "Is it absolutely perfect? How can it be improved? If it was improved, who would benefit from it being improved? How difficult would it be to improve it? Should I take action to get it improved?" During this process, improvement opportunities are identified and evaluated to determine how the organization or its stakeholders could benefit from improving them. It will also include doing a preliminary rough analysis of what impact successfully completing the project would have on the value-added content received for your stakeholders. Should it be related to a new or current process, a project, or a product? These potential opportunities are like puffs of smoke that float around us all the time and can disappear from our thoughts as quickly as a strong wind blows a puff of smoke away. Every place we look we see opportunities we could jump on but are quickly pushed aside by other opportunities. The real problem we have is not identifying the opportunities; it is being able to hold onto them and determine if this is the right opportunity for you to champion. We believe that none of us has the ability to take on all the improvement opportunities we identify every day. One of the longest lists I have on my computer is my list of improvement opportunities that are available to me and I would like to take advantage of but do not have the time or resources to do right now.

A truly observant person soon becomes overwhelmed with the number of new or improvement opportunities that are available to each of them. Sometimes these opportunities are personal, like a pretty girl who sits in the second seat in the front row. Sometimes it is family related, like building a mother-in-law's house out back and moving grandma into it so she would have her own place. Other times it's new product related, like we could use this sticky mess we are cleaning up as a good water-resistant epoxy. It could be as small as moving the bin of screws to the left-hand side of the workbench, making it much easier and faster to put an assembly together. We categorize these opportunities as follows:

1. Work product-related opportunities
2. Work process-related opportunities
3. Service-related opportunities
4. Self-related opportunities
5. Family-related opportunities
6. Personal-growth opportunities
7. No value-add opportunities
8. Self-implementation opportunities
9. Opportunities that are not within my personal or organizational principles
10. Opportunities that are forced upon me
11. Opportunities where someone else should take the lead
12. Impossible (extremely difficult and time-consuming)

13. Opportunities where the return on investment is not adequate
14. Not part of the organization's mission

(Note: You may want to add additional headings or remove some of the 14 headings that we presented.)

In this book, we are going to be focused on business-related new or improvement opportunities. These are opportunities that would generate added value to some or all of our stakeholders. To get the process started, we could identify those activities that each function has related to the PIC. The functions that generate the biggest value-added activities usually are research and development, product engineering, marketing, and sales. At a very minimum, you need to identify the roles, responsibilities, and accountability for these four main functions.

An organization's reputation as an innovative organization is largely based upon how creative and unique its interface with the customer or potential customers are. We asked random customers who they consider as innovative organizations. Then we asked the question, "Why do you consider these organizations as being innovative?" Over 95% of the time their opinion was based upon the innovation or uniqueness of the organization's output or was based upon what someone else had told them (friends, TV, magazines, etc.). It is not the IMS that Apple has developed that has earned them the reputation as one of the most innovative companies in the world; it is the product that they have delivered. This is when you should ask the questions: How well does the output function? What unique features does it have? How easy is it to use? Take a minute to see if you don't agree that most of the time it is the uniqueness of your output that gives an organization a reputation as being innovative.

Make a list of 10 organizations that you consider to be innovative. Then honestly define why you placed each organization on the list. I'll give 5 to 1 odds that your conclusion will be based on your view of the organization's outputs. (In defining "output," we include advertising and sales.) It's not how good your product is; it's how you make people think it is either through actual use or from advertisements. I drive a Buick with over 200,000 miles on, and the only problem I've had is that every 60,000 miles I have to buy a new set of tires. And I had to fix the automatic window on the driver side. But Consumer's Report tells me that Toyota is a better car with fewer breakdowns and more reliable, so I have a tendency to believe that Toyota is a more innovative company.

There are some organizations that obtain a reputation for being innovative not based upon the actual product that they deliver but based upon their promotions (television, radio, Internet, magazines, conferences, etc.). They repeatedly announce that they are innovative until you begin to believe it. Keeping this in mind, we are going to focus this part of the book on how you process innovative ideas that are designed to be used by the general public or a specific party. We also will focus upon the part of the organization that is most involved in the task related to how you should use your IMS. Remember your product design is only one part of delivering an innovative output to your internal and external customers.

Inputs to Process Grouping 1: Opportunity Identification

There are many ways to identify potential improvement opportunities. Please note that throughout this book, we are using innovation opportunities and improvement opportunities interchangeably. When you look for opportunities, you find both innovation and improvement opportunities. We probably should refer to it as identifying innovation and improvement opportunities.

The following are 10 typical popular ways to identify improvement opportunities.

1. Customer suggestions
2. Technology advancements
3. Employee suggestions
4. Competition's new products
5. R&D experiments
6. Technical journals
7. Professional societies
8. Focus Groups with customers
9. Supplier suggestions
10. Personal observations

You probably can add an additional 10 easily. Each of these ways to identify innovation opportunities is supported by its own process. Within the same organization, it is frequently the case that different functions have design processes that are unique to their specific needs and preferences.

With hundreds of opportunities that are available to most organizations, it is important that they select the right opportunities to continue to invest resources in. This normally requires an excellent understanding of the type of business the organization is in and what is happening to this technology and how their competitors are progressing. The two main ways to obtain this type of information is through information documented in the public domain.

The two main ways to obtain data are retrieving published data (completed research) that is in the public domain (books, magazine articles, conferences, technical reports, etc.) and conducting original research (observing, interviews, testing/disassembling competitive products, and services). All of these methods provide excellent input related to identifying an improvement opportunity and the development of solutions.

The following are some of the most popular inputs to Process Grouping 1. Opportunity Identification to help identify innovation opportunities.

1. Opening your eyes and senses
2. Opening your mind
3. Customer suggestion
4. Surveys
5. Focus groups
6. Marketing opinions
7. Sales opinions
8. Technology improvements
9. Customer complaints
10. Performance problems
11. Negative return on investment
12. Improve competitor product features and performance
13. Benchmarking studies

My personal favorite way of identifying improvement opportunities is simply by sitting down with a pencil and sheet of paper and writing down the things that I'm unhappy with or I would like to see changed. I find that closing my eyes in this process helps. But of course, I open my eyes to make my list. I then check off the ones that I would like to be involved in changing and then check off the ones that I feel it is worth my time to get changed. Next, I check off the ones that would be value added to the organization where I work. The items that had three checks behind them are very good candidates for my personal improvement opportunities list.

I then look around where I am sitting and make a list of the items that could be improved. For example, the hooks on the drapes are difficult to keep on the drapery rod. There must be a better design for them. This list I process in the same manner I did my personal improvement opportunities list.

For the next two days, I'm looking for things in the environment that could be improved or should be changed. I add this to my already long list and repeat the rating process again. Without exception every time I do this, I end up with far more things I want to do than I ever will have time to do. Now the hard part is looking at the list and selecting the one or two things I'm going to start working on today.

There are hundreds and hundreds of improvement opportunities just waiting for you to identify; some of them are personal in nature, and others are directly related to the organization you work for. Some of them you think you can come up with a better item, and others you think you cannot improve. If you think you cannot do it, you will be right. I'm always surprised at what I can do if I try to do it. Just because you never did it before is a very good reason why you should try to do it now.

Of course, the old tried and proven ways of identifying improvement opportunities are very beneficial – like listening to customer complaints, asking customers what they want and what they think they want in the future, staying abreast of technology changes that could impact your product design, analyzing costly activities, benchmarking, and reengineering.

Some of the processes/activities that make up Process Grouping 1. Opportunity Identification are:

1. Motivating individuals to recognize opportunities
2. Defining process alternatives for a specific project/program
3. Preparing to use the process
4. Using the process
5. Identifying potential opportunities
6. Evaluating potential opportunities
7. Doing a rough value-analysis for high value added opportunities
8. Management approval for further refinement of opportunities
9. Obtaining management approval of mission statements for every approved opportunity

A good class focused on how to identify improvement opportunities will quickly make a major change in the way you look at things. Every place you look, everything you read, everyone you talk to, almost anything you do, is going to all of a sudden present opportunities that you've never realized before. We recommend that you make a list of improvement opportunities that you identify. You will be surprised at how quickly the short list becomes a very long list, and you start to get your red pencil out to cross out ones that you don't want to consider.

A typical list contains any personal items, family items, your organization's business items, other organization's improvement opportunities, and ideas to build your new business around and make changes in supplier inputs, etc. These improvement ideas can vary all the way from buying a helicopter to fly to work so you can eliminate all the traffic on the freeway, to combining the state government with the federal government to eliminate the state government to save money and make faster decisions. It could also include something like the ridiculous idea of going to Hollywood to marry a movie star so she could support you (the sublime). Or rearranging your office so you can get another file cabinet, relocating the coffee machine so that is closer to your desk, allowing everyone to have their own coffee maker on their desk, developing a substitute for salt that has no sodium in it, buying the Dragon software package for everyone in the company so no one would need to be able to type, opening a steakhouse where the center of attraction would be a campfire with a group of cowboys sitting around singing Western songs, developing an ink that is erasable, relocating the handle on our newest product so that it would be easier to move and not skin your knuckles, moving the drilling operations to before the grinding operations so that the drilling burs would be removed during the grinding operation, developing an unbreakable screen of the cell phone, developing a machine that is light enough to sit on an individual's head that will help old people keep their balance, eliminating the need for canes, holding executive training sessions in Alaska at a fishing camp, fishing for four hours a day and doing organizational work for 10 hours per day, developing a company that organizations could outsource their entire quality program to, developing a standard home toaster that would toast the bread equally across the slice, developing a simple way that could be used to edit webinar recording, reorganizing the organization by breaking it up into small business units that are self-funding and totally accountable, starting providing real people contact for customers when customers call in for help or to complain, eliminating the four hours on hold when you're trying to reach the company's complaint department, or combining departments 3, 6, and 8 together.

Enough of the random improvement opportunities examples. The point is that improvement opportunities can vary all the way from the ridiculous to the sublime. Many of the improvement opportunities will not be related to the business that the organization is involved in. It is for that reason, one of the very first things we need to do in developing an organizational innovative management system is to separate the improvement opportunities that our employees define that are not related to the business from the ones that are going to drive future value added from the organizational standpoint and its stakeholders.

Furthermore, there is even a second important consideration that must be looked at early in the organizational innovative management cycle. It is the improvement opportunities that are already budgeted for and included in the day-to-day operations within the organization. These are day-to-day activities that are covered in the individual's job description and budgeted for in the normal budgeting cycle. Typically, these are problems that need to be addressed without requiring additional budgetary resources and improvements or refinements to the present operating systems that are minor in nature. Example: Developing repair routings, updating procedures, developing new commercials, training for new employees, updating the operating manual, selecting new suppliers preparing annual reports, etc.

The combination of separating out improvement opportunities that will not result in direct value added to the organization's overall performance and removing improvement opportunities that are covered in the normal budgeting cycle typically will illuminate between 60% and 90% of all the improvement opportunities generated by first-level managers and employees. Typically, it will account for 40% to 70% of the improvement opportunities identified by middle-management and the executive team. It is absolutely essential that we separate these improvement opportunities out of all the organizations' Innovative Management System very early in the cycle. This should take place during Process Grouping 1. Opportunity Identification.

Process Grouping 1: Opportunity Identification Activity Block Diagram

There are three major work types of Innovation Opportunities:

1. Product Innovation Opportunities – These are opportunities that result in delivering new or improved products.

2. Process Innovation Opportunities – These are opportunities that result in delivering new, improved processes that are more efficient, effective, and adaptable.
3. Service Innovation Opportunities – These are opportunities that relate to the service industry where the primary considerations are the interface between the consumer and the individual delivering the service (e.g., banks, insurance companies, grocery stores, or consulting firms).

In order to keep the size of this book to a reasonable length, we decided to present Activity Block Diagrams in this and the next two chapters based only on the *product* innovation opportunities, as it is the most complex of the three opportunity classifications. For those of you that would like to apply this methodology to process redesigning or to the service industries like healthcare, banking, and entertainment, we recommend that you make a list of all of the activities shown in the Activity Block Diagrams in the three chapters and then evaluate each block to determine if it applies to your organization or the project/program you are responsible for. In most cases, you'll find that there are a number of activities required in the product cycle that aren't required in either the process redesign or service product delivery systems and very few additional ones that need to be added.

As discussed in Chapter 5, ISO 56002:2019 suggests that the organization's innovation processes be modified to suit the individual initiative depending upon the type of innovation and the circumstances within the organization. The *Enhance* cloud service at **EDGESoftware.cloud/managing-innovation** supports this by allowing you to start with the Best Practice processes you are learning about in this book and then customize them for your particular needs.

It is important to understand that the Activity Block Diagram for an opportunity will differ based upon the type of opportunity that is being pursued. The three major types of Activity Block Diagrams are:

- Product Activity Block Diagram
- Service Activity Block Diagram
- Process Activity Block Diagram

In the following Product Activity Block Diagram, initially, the team was looking at opportunities and classified them into one of the three main categories. As we look at the innovation cycle for these three innovation categories, we find that the three high-level phases are common to each of the three categories (Phase 1. Creation, Phase 2. Preparation and Producing, and Phase 3. Delivery). Even the 12 Process Groupings are similar in each of the three categories, but the processes within the Process Groupings have a great deal of variation from category to category, and as a result, the innovation cycle for each of the three categories is significantly different. It is easy to separate product-driven organizations from service-driven organizations. This provides the first general guidance in determining if you should be working on products or service-type innovation opportunities. This certainly does not close out the need for innovation in all these categories.

Certainly process innovation improvement is an important driver in all organizations and is particularly critical in not-for-profit and governmental organizations. At that point in the cycle service opportunities were separated and set aside for Service Activity Block Diagram. The innovative opportunities that are classified as process related are also separated and set aside for the Process Activity Block Diagram.

For brevity's sake, we have only provided you with a detailed Activity Block Diagram for product-related innovative opportunities.

Typical activities that are included in Opportunity Identification (see Figure 6.1).

1.1. Identify potential innovative opportunities.
1.2. Is it part of the organization's mission?
1.3. No – Terminate activities within the organization.
1.4. Classify potential Innovative Opportunities into three categories (Service, Process, Product).
1.5. Classified as service innovative opportunity.
1.6. Classified as a process innovative opportunity.
1.7. Classified as a product innovative opportunity.
1.8. Will it meet initial screening activity?
1.9. No – Drop from consideration.
1.10. Yes – Conduct Tollgate I.
1.11. Did it meet value-analysis requirements?
1.12. No – Drop from consideration.
1.13. Yes – Prepare a preliminary project charter.
1.14. Set up a knowledge warehouse.

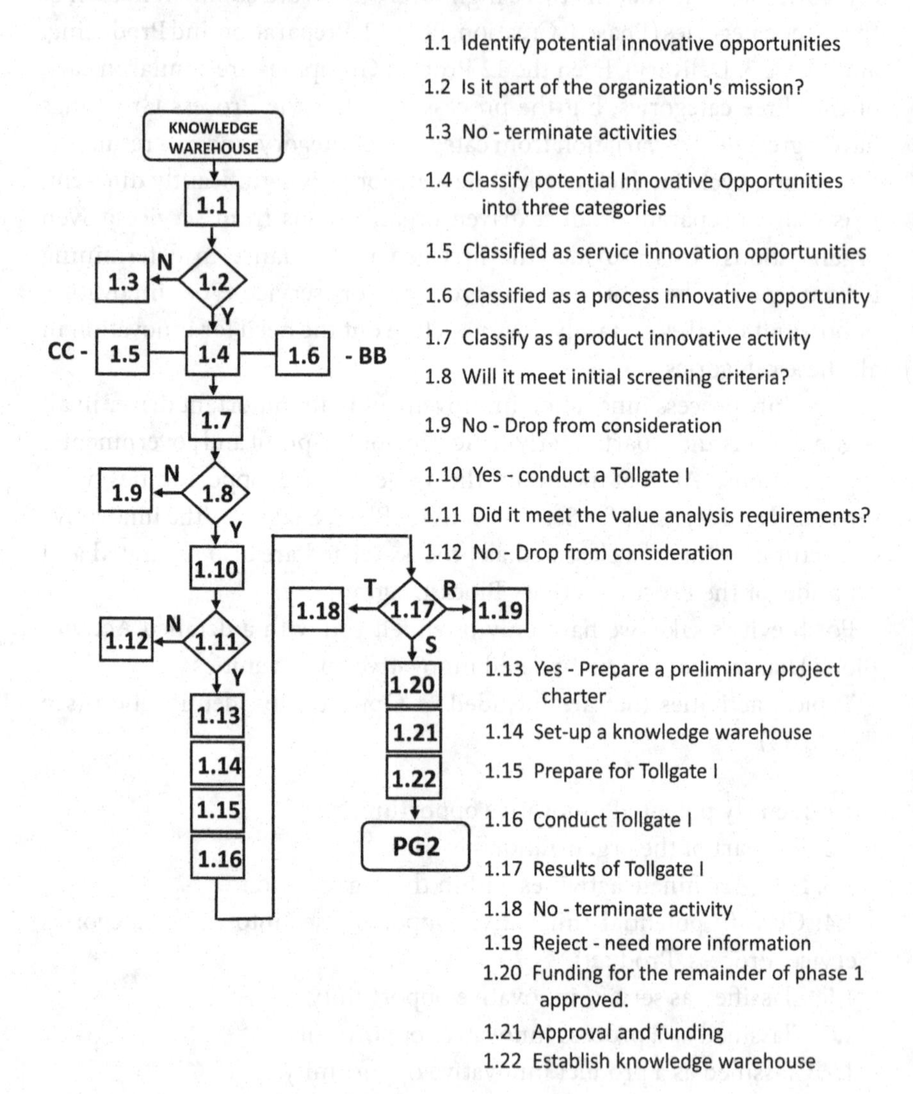
PHASE I: Process Grouping 1 – Opportunity Identification and Tollgate I – Opportunity Analysis (ABD)
1.1 Identify potential innovative opportunities
1.2 Is it part of the organization's mission?
1.3 No - terminate activities
1.4 Classify potential Innovative Opportunities into three categories
1.5 Classified as service innovation opportunities
1.6 Classified as a process innovative opportunity
1.7 Classify as a product innovative activity
1.8 Will it meet initial screening criteria?
1.9 No - Drop from consideration
1.10 Yes - conduct a Tollgate I
1.11 Did it meet the value analysis requirements?
1.12 No - Drop from consideration
1.13 Yes - Prepare a preliminary project charter
1.14 Set-up a knowledge warehouse
1.15 Prepare for Tollgate I
1.16 Conduct Tollgate I
1.17 Results of Tollgate I
1.18 No - terminate activity
1.19 Reject - need more information
1.20 Funding for the remainder of phase 1 approved.
1.21 Approval and funding
1.22 Establish knowledge warehouse
KNOWLEDGE WAREHOUSE
1.1
1.2
1.3
N
Y
CC -
1.5
1.4
1.6
- BB
1.7
1.8
1.9
1.10
1.11
1.12
1.13
1.14
1.15
1.16
T
R
S
1.17
1.18
1.19
1.20
1.21
1.22
PG2

FIGURE 6.1
Process Grouping 1. Opportunity Identification and Tollgate I Activity Block Diagram

1.15. Prepare for Tollgate I.
1.16. Conduct Tollgate I.
1.17. Results of Tollgate I.
1.18. No – Terminate activity.
1.19. Reject – Need more information.

1.20. Funding for the remainder of Phase I approved.
1.21. Approval and funding.
1.22. Establish knowledge warehouse.

Output – forward to PG2. Opportunity Development.

This is where an individual or group view the same old situation and see it in a different light than had been viewed before. It's where an individual or group states, "We should be able to do it differently bringing additional value to the organization." At this point, the individual or group usually do not know how to make the improvement, but they are committed to coming up with an innovative/creative solution. Frequently this ends up with a mission statement being approved.

TOLLGATE I

Tollgate I is a high-risk Tollgate. All that management has available to make a decision to go forward is a potential improvement/innovation opportunity that an employee and a minimum of one manager is interested in investing risk resources to determine if it can be developed to take advantage of this opportunity.

Decisions at this Tollgate I level are often highly based upon management judgment and past experience. Basically, Tollgate I approval provides resources for a group of one or more individuals to investigate the improvement opportunity and determine if the organization should take advantage of it and if it would result in adding value to the organization and/or its stakeholders.

The sponsors of the improvement opportunity need to provide the following information in order for the decision to be made related to continuing work on taking advantage of the improvement opportunity.

1. A description why/how the proposed activity is an improvement opportunity.
2. What would need to change to take advantage of the improvement opportunity?
3. What would be the impact on each of the stakeholders if the opportunity is taken advantage of?

4. Does the sponsoring team have any ideas on how to take advantage of the opportunity?
5. How much resources are required to advance the project to Tollgate II? Who will need to supply these resources, and are these resources available at the present time?
6. What is a general timeline of how long it will take to advance the program through the value proposition approval activity?
7. What is the potential value added from the opportunity if it is implemented?
8. Are there any estimates of what the implementation costs could be?
9. Who would be responsible for managing the activities and held accountable for the results?

In conducting Tollgate 1. Opportunity Analysis, a very rough draft of return on investment analysis is prepared. The first part of that analysis is determining what should be changed. The next part is to analyze what is the one-year added-value content that is generated if it is changed. We like to develop the maximum value added that the change could result in and then estimate conservatively how effective the change would be. For this we assume the change will be between 50% and 75% effective. Very few changes are 100% effective with no negative impacts. Now make an evaluation if it is worthwhile pursuing the opportunity any further.

If the decision is made to continue, make an estimate of the dollar amount of resources (employee time, materials, equipment, etc.) required to investigate in developing a way to take advantage of the opportunity. The employee time should include the individual, team members, management reviews, support service personnel, assembly personnel, etc. Once you have estimated the total dollar amount of resources required to develop a recommended action, then double that value, because we find these estimates usually are off by 80% to 200%.

You are now in a position to make a very rough estimate of the percentage return on investment, even though you haven't estimated how much it will cost to implement the change because the change has not been identified at this point in time. As a result, we question going forward with any evaluated opportunity that doesn't calculate out at a minimum of 400% return on investment based upon the costs of investigating the opportunity and preparing an action plan.

The exception to this is improvement opportunities that are related to safety; in these cases the risk analysis needs to be used to determine if the opportunity will be dropped or continued.

Between 70% and 80% of all the initiatives will require only the immediate manager's approval to progress through to Tollgate II in the cycle. These are usually situations where the individual recognizing the opportunity or rearranging assignments within the department is able to develop an approach for taking advantage of the opportunity and prepare a value proposition that will be presented at Tollgate II. Remember that 95% of all patents are obvious evolutionary changes and show very minor changes.

Usually when resources outside of the department are required to develop a way to take advantage of the opportunity and prepare a legitimate value proposition, the decision would be made by a group of affected managers. Opportunities where a major investment will be made, which is not part of the normal job assignment of already-budgeted activities, should be made by upper management and reflected in the departmental budgets. It is estimated that this would be less than 5% of the opportunities identified.

PROCESS GROUPING 2: OPPORTUNITY DEVELOPMENT

At this point, you define ways to take advantage of the opportunity and then assign individuals or a team to look at the many different ways the opportunity can be addressed. It calls upon the individual to step out away from his or her daily activities and use their mental capabilities to come up with new and unique solutions. There is a great deal of self-satisfaction and pride when the individual or team defines one or more ways to take advantage of an opportunity that no one else was able to take advantage of.

Due to the high degree of innovation, the complexity of opportunity development, and the scope of the type of outputs that will be involved, we have divided this section up into three Activity Block Diagrams that will be presented.

1. Outputs that would be classified as an apparent solution or minor innovations. Based upon reviews and classifications of tens of thousands of patents, this accounts for over 95% of the patents issued.

2. Outputs that would be classified as major innovations. This accounts for just over 4% of the patents.
3. Outputs that would be classified as new paradigms or discoveries. This accounts for 0.3% of the patents issued.

As we begin Process Grouping. Opportunity Development, we have a mission statement and approval to form a team that will be responsible for collecting sufficient data so that they can develop one to four potential ways to take advantage of the assigned opportunity. Also, a group leader and a project sponsor have been assigned to be responsible for coordinating the team's activity and keeping the project on schedule. Included in the mission statement was the minimum value-added return on investment that was required and the maximum cycle time that would be devoted to coming up with the answers and preparing a value proposition. It is during Process Grouping. Opportunity Development where the magic occurs and creativity is the most valued trait that the team members could have. The data collected in Process Grouping. Opportunity Identification is also made available, but it usually is not sufficient to define the root cause that created the opportunity.

Objective of Process Grouping 2: Opportunity Development

The objective of this process is to develop a minimum of two and preferably three or four options that could be implemented to take advantage of the defined opportunity.

An Innovation Project Team (IPT) is often formed with the responsibility for collecting the information and knowledge that will allow them to creatively meet the assigned objectives. Although this team has a very narrow objective, the participants usually continue with the project through Phase I. Creation and Phase II. Preparation and Producing if the project is approved to be added to the organization's portfolio of active projects.

After an IPT has been formed, it is trained to use the necessary innovation tools and methodologies. Some of their activities include the following:

1. Define how success could be measured, how to measure it, measure current status, and set goals for value-added content. These goals should be in line with the goals defined in the approved objectives.

2. Develop a team mission statement and team charter that is approved by the appropriate personnel.
3. Brainstorm to develop a list of what needs to be done in order for the team to meet their objectives.
4. Define and understand the knowledge areas that could influence the suggested changes.
5. Evaluate to determine if organizational change management concepts need to be formally applied, and if so, how and when?
6. Develop their data collection plan and implement it after they have the approval of the managers whose employees will be affected. They must be sure to collect a large enough sample to convince yourself and management that they feel comfortable making decisions based upon the data presented. Be sure that the team has one-on-one discussions with the individuals who will be affected.
7. While the data is being collected, we find it is a good time to do a strengths/weaknesses analysis of the opportunity. Remember for every positive thing that is done, there is a negative thing that will occur that can offset the positive action.
8. Don't wait until all the data is collected before you start analyzing it. Too often you will be surprised that the data being collected was not what you thought it would be. By analyzing the data as it comes in, you can recognize the problem early in the cycle and it will allow you time to collect a different set of data without delaying the project. As the data comes in, it should be validated and entered into the IPT Knowledge Bank.
9. Using the information in the knowledge bank, brainstorming the team's own experience and creative ideas that come up during the team meeting, it should provide the IPT enough knowledge to identify the factors that are restricting the opportunity from performing better (identify root causes).
10. Once the restricting factors are identified, tools and methodologies similar to the 76 tools listed in Appendix B will enable the team to develop creative and innovative ways to take advantage of the opportunity.
11. The IPT typically uses a lawyer to review the proposed action plans and determines what needs to be done to not infringe on another organization's patents/copyrights and to protect the knowledge

VALUE ADDED	Low	RESOURCES REQUIRED High
High	$V_H - R_L$ Very Good	$V_H - R_H$ Maybe
Low	$V_L - R_L$ Maybe	$V_L - R_L$ Bad

FIGURE 6.2
Opportunity Selection Box

assets that the organization has. The earlier you can do this in the cycle, the better.

12. Once IPT defines one potential action plan, they should immediately turn around and use a different set of assumptions to create second and third alternative solutions. This often drives to a fourth alternative that takes advantage of the best innovative ideas in the three previous ones.
13. Using the Opportunity Selection Box to select potential innovation opportunities, an opportunity that is located in the upper left-hand quadrant indicates value added is high and the resources required to take advantage of the opportunity is low (see Figure 6.2). An opportunity classified as having this combination is usually a very good innovation opportunity. For an opportunity located in the lower right-hand quadrant, it indicates low value added and will require a large resource investment in order to take advantage of the opportunity. This is normally a bad situation. The last two quadrants are classified as "Maybe" and require more discussion before a decision is made to continue or table the proposed opportunity. Extra consideration is often given to opportunities that have a direct positive impact upon the organization's customers/consumers.

This may look like a very simple way of classifying innovation opportunities, and it is if you keep the estimate considerations very simple and straightforward. Using the same system at the business case analysis process grouping level requires a great deal of more thought and consideration related to what value added is. In most estimates that we've seen, they only talk about upfront tangible savings, ignoring the negative impacts that the opportunity may create and often not considering its impact upon all of the stakeholders. For example, what is the value added of the change in design that eliminates the need for two employees at a salary level plus variable overhead of $25,000 per year. Most estimates would say the value-added content would be a savings of $50,000. And that is a real value added through the organization. But what is the negative value added to the employee who is now without a job, and what is that employee's negative impact upon the economy? Some of the very advanced companies are now considering total impact on all stakeholders in their decision-making processes, but we are estimating that is less than 0.1% of the organizations in the United States. It certainly is a lot simpler when they can limit our thinking to impact upon the organization ignoring the rest of the stakeholders.

Process Grouping 2: Opportunity Development Activity Block Diagram for Apparent and/or Minor Opportunities

Typical activities that are included in Opportunity Development for Apparent or Minor Opportunities are defined in the Activity Block Diagram in Figure 6.3.

2.1. Form Opportunity Development Team.
2.2. Develop team charter, project objectives, innovation goals, and operating ground rules.
2.3. Develop data collection plan.
2.4. Implement data collection plan.
2.5. Analyze data to define root cause and/or improvement opportunities.
2.6. Develop an action plan to meet the teams and innovation goals.
2.7. Define other approaches that could be used.
2.8. Select the best two approaches.
2.9. Input into knowledge warehouse.
2.10. Develop action/implementation plan.
2.11. Update knowledge warehouse.

During this activity, a number of potential problem solutions or improvement opportunity will be identified, analyzed, and prioritized. Also, during this activity, steps are actually taken to protect intellectual capital (patent new and unique concepts or chart to see that there are no patent infringements).

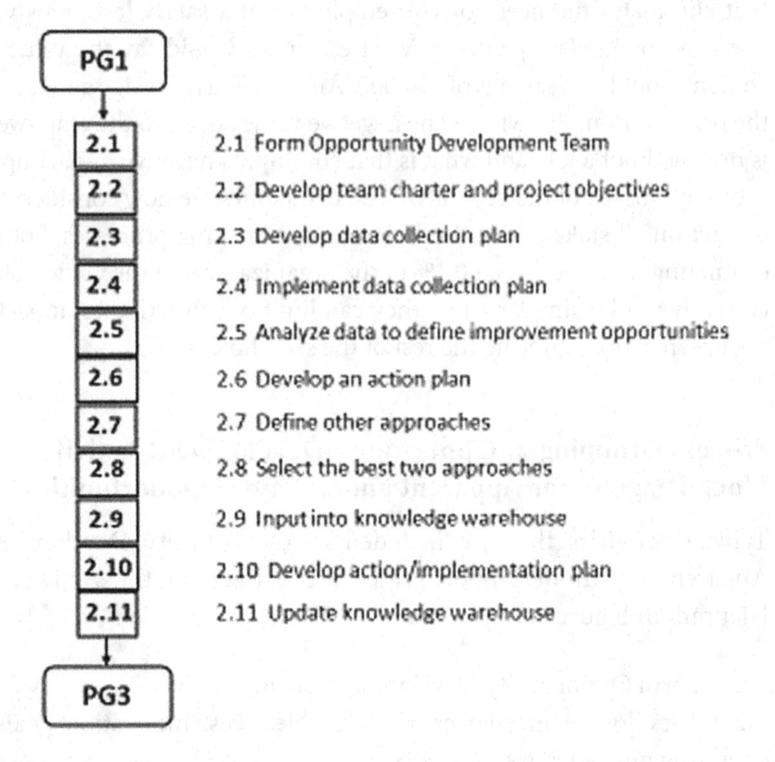
PHASE I: Process Grouping 2 – Opportunity Development for Problem Solving Process for Apparent and/or Minor Opportunity (ABD)
PG1
2.1
2.2
2.3
2.4
2.5
2.6
2.7
2.8
2.9
2.10
2.11
PG3
2.1 Form Opportunity Development Team
2.2 Develop team charter and project objectives
2.3 Develop data collection plan
2.4 Implement data collection plan
2.5 Analyze data to define improvement opportunities
2.6 Develop an action plan
2.7 Define other approaches
2.8 Select the best two approaches
2.9 Input into knowledge warehouse
2.10 Develop action/implementation plan
2.11 Update knowledge warehouse

FIGURE 6.3
Process Grouping 2. Opportunity Development Activity Block Diagram for Minor Opportunities

Process Grouping 2: Opportunity Development Activity Block Diagram for Major Opportunities

Typical activities that are included in Opportunity Development for Major Opportunities are defined in the Activity Block Diagram in Figure 6.4.

Note: From this Activity Block Diagram for Process Grouping 2, we have not included the details related to each block on the Activity Block Diagram due to the book page-size limitations and to make it easier for you to read.

2.30. New design is the desired opportunity.

2.31. Form an Innovative Opportunity Development Team.

2.32. Define performance/physical desired changes.

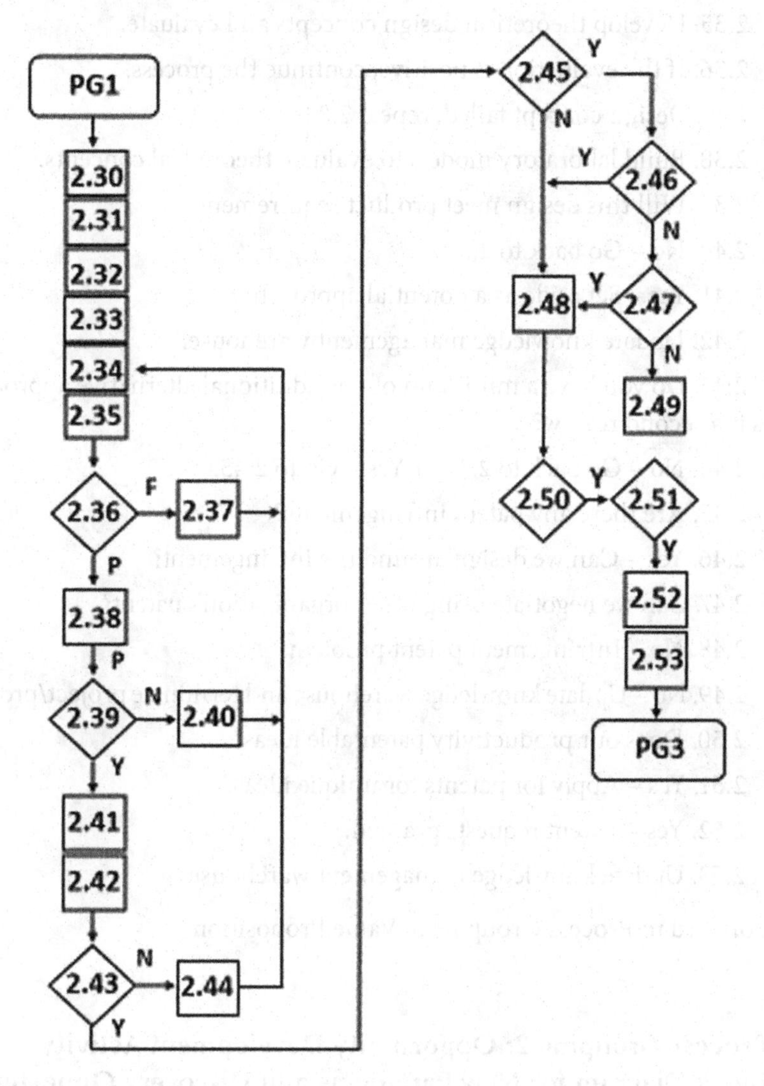

FIGURE 6.4
Process Grouping 2. Opportunity Development for Major Opportunities Activity Block Diagram

2.33. Set magnitude of desired changes.

2.34. Define technology changes from present design.

2.35. Develop theoretical design concepts and evaluate.

2.36. If the evaluation is positive, continue the process.

2.37. Design concept failed, repeat 2.34.

2.38. Build laboratory models to evaluate theoretical concepts.

2.39. Will this design meet product requirements?

2.40. No – Go back to 2.34.

2.41. Yes – Set aside as a potential approach.

2.42. Update knowledge management warehouse.

2.43. Do you have a minimum of one additional alternative approach or is it a second review?

2.44. No – Go back to 2.34 or Yes – Go to 2.45.

2.45. Are there any patent infringements?

2.46. Yes – Can we design around the infringement?

2.47. Can we negotiate using other organization's patent?

2.48. No – Infringement patent problem.

2.49. No – Update knowledge warehouse and terminate project/program.

2.50. Does our productivity patentable ideas?

2.51. Yes – Apply for patents for unique ideas.

2.52. Yes – Patent requests granted.

2.53. Update knowledge management warehouse.

Forward to Process Grouping 3. Value Proposition.

Process Grouping 2: Opportunity Development Activity Block Diagram for New Paradigms and Discovery Opportunities

Less than 0.06% of items classified as innovative can be classified as unique new discoveries and 0.24% in new paradigms. The majority of the organizations using an IMS will ever produce an output that can truly be classified as a discovery. Most innovative outputs or processes are developed based upon previously known and understood concepts that are improved upon or slightly adapted to a new situation. Discovery classified innovations often occur as a surprise or from the result of something else that is being developed. In other cases, a concept is accepted by a research group to evaluate and determine if it will lead to something new and different from what we do not have today. These discovery innovations are the result of things like:

- I forgot to put it in the refrigerator last night and taste how good it is.
- I mixed the wrong chemicals together and see how sticky it is.
- I forgot to turn off the electricity and look how bright it glows.
- The odor in the room got so strong I couldn't stand it, and all of a sudden, my migraine headache went away.
- I ran over it with the truck, and it didn't even dent the can.

All of these examples could be classified as unexpected or error-related results. Some of these discoveries are related to targeted opportunities. Targeted opportunities are opportunities that have been identified but will require the creation of a new and unique method or approach in order to take advantage of the opportunity. Frequently these are related to healthcare situations. For example, some of the targeted opportunities that we have today:

- Develop a cure for Alzheimer's disease.
- Develop a cure for old people losing their balance.
- Define a worldwide management system that will eliminate wars.
- Develop a system that will eliminate the need for electrical wiring within the house.
- Develop a system that will transfer information directly into the brain without using any of the senses.

Typical activities that are included in Opportunity Development for New Paradigms and Discovery Opportunities are defined in the Activity Block Diagram in Figure 6.5.

2.60. Targeted opportunities.
2.61. Unexpected results or error results.
2.62. Establish project and budget to evaluate the unexpected results.

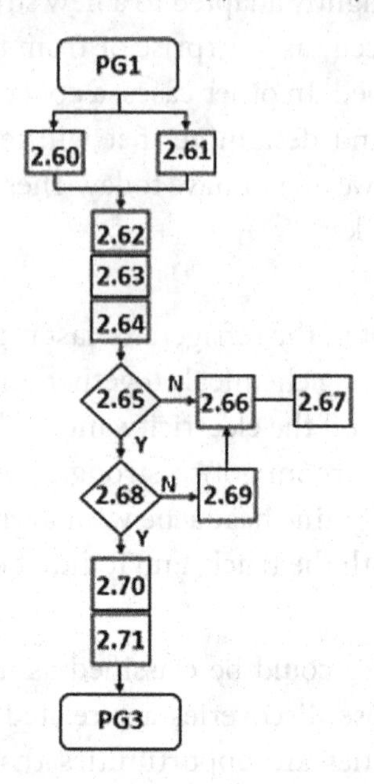

FIGURE 6.5
Process Grouping 2. Opportunity Development for New Paradigms and Discovery Opportunities

2.63. Experiments to understand the results.
2.64. Literary research to explain the results.
2.65. Are the results understood?
2.66. No – Update the knowledge management warehouse.
2.67. Terminate project.
2.68. Yes – Define potential applications.
2.69. No applications defined – Update the knowledge management warehouse.
2.70. Yes – Needs further development.
2.71. Put on hold until a need is defined.

Targeted opportunities are often long-range, high risk, and very expensive. They often require a great deal of capital equipment and human resources. As a result, very few companies can afford to invest heavily into basic research, and they rely heavily on the government and universities to develop the basic concepts and applications.

Once you have developed a potential solution, the next question you have to ask yourself, "Is this potential solution value added to the stakeholders?" If you come up with a number of potential solutions, you need to evaluate which solution creates the most value added per investment. Again, this Process Grouping requires a great deal of imagination, estimation, and research. It relies heavily on the judgment of the individual defining the potential solution. This process evaluates the creativity of the individuals developing the potential solution. Failure at this point in the system often results in reevaluating of the opportunity to see if it should be terminated or, at a minimum, in sending it back to develop a new and more comprehensive potential solution.

The primary outputs from Process Grouping 2. Opportunity Development are as follows:

a) Mission statement
b) Successful completion of Tollgate I
c) Resources approved up to Tollgate II
d) Normal functional improvements opportunities classified and scheduled
e) High-value opportunities identified with management agreement
f) Knowledge management system updated
g) Potential innovation opportunities identified and roughly qualified
h) Start of a database related to the project

PROCESS GROUPING 3: VALUE PROPOSITION

A value proposition is a document that defines the benefits that will result from the implementation of a change or the use of an output as viewed by one or more of the organization's stakeholders. A value proposition can apply to an entire organization, parts thereof, or customers, or products, or services, or internal processes. It usually does not consider other ways that the resources could be used to get a higher value added. This type of consideration will be handled during Process Grouping 5. Business Case Analysis.

Inputs to Process Grouping 3: Value Proposition

Some of the inputs to Process Grouping 3. Value Proposition include an updated knowledge management system, a project mission statement, organization values and principles, the opportunity development team's records and engineering notebooks and knowledge protection results (Source: *Maximizing Value Propositions to Increase Project Success Rate*, H. James Harrington and Brett Trusko). See Figure 6.6.

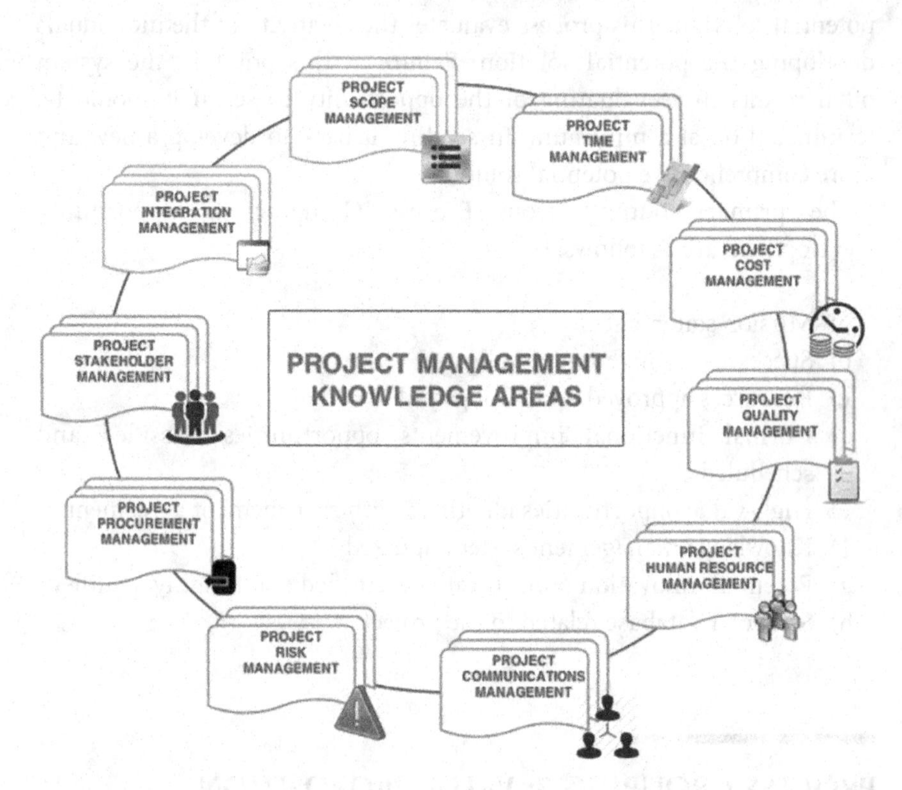

FIGURE 6.6
The Project Management Knowledge Areas as Defined by the Project Management Institute Located in Canada

1. Project Integration Management
    - Develop Project Charter
    - Develop a Preliminary Project Scope Statement
    - Develop a Project Management Plan
    - Direct and Manage Project Execution
    - Monitor and Control Project Work
    - Integrate Change Control
2. Project Scope Management
    - Scope Planning
    - Scope Definition
    - Work Breakdown Structure
    - Scope Verification
    - Scope Control
3. Project Time Management
    - Activity Definition
    - Activity Sequencing
    - Activity Resource Estimating
    - Schedule Development
    - Schedule Control: Project Tracking
4. Project Cost Management
    - Cost Estimating
    - Cost Budgeting
    - Cost Control
5. Project Quality Management
    - Quality Planning
    - Perform Quality Assurance
    - Perform Quality Control
6. Project Human Resource Management
    - Human Resource Planning
    - Acquire Project Team
    - Develop Project Team
    - Manage Project Team
7. Project Communications Management
    - Communications Planning
    - Information Distribution
    - Performance Reporting
    - Manage Stakeholders

8. Project Risk Management
    - Risk Management Planning
    - Risk Identification
    - Qualitative Risk Analysis
    - Quantitative Risk Analysis
    - Risk Monitoring and Control
9. Project Procurement Management
    - Plan Purchases and Acquisitions
    - Plan Contracting
    - Request Seller Responses
    - Select Sellers
    - Contract Administration
    - Contract Closure

Usually the team that developed the approaches that the organization will use to take advantage of the opportunity will develop and present the value proposition to management. For improvement opportunities that may become part of the organization's portfolio of active projects, at least one member of the team should be well acquainted with the Project Management methodologies. They should also know how to prepare a project plan in keeping with the PMBOK methodologies. The Project Management Institute located in Canada defined nine management knowledge areas that project teams need to be skilled in. Recently, a tenth project management knowledge area has been added. It is stakeholder management.

Typically, the value proposition is prepared by the team that developed the approaches to take advantage of the improvement opportunity (Process Grouping 2. Opportunity Development). For projects that have the potential of becoming part of the organization's portfolio of projects, at least one member of the team should be very familiar with the PMBOK created by the Project Management Institute.

Although the term "value proposition" is used freely in the modern business lexicon, the truths about value propositions are that (1) they are generally not well thought out and executed, and (2) the process one follows to prepare a quality value proposition is generally misunderstood. The value proposition is the specific, quantified, and identified opportunity to improve business results. In the discipline of innovation, they are the justification for action or inaction. They are an opportunity for the innovator who may be intimidated by the "organization" to be able to prepare a document outlining the value of an idea, process modification, or hunch. The creation of an opportunity center (innovation center, creativity center, etc.) is an effective mechanism to move the idea from the rank and file to the organization. The value proposition is the mechanism for identifying the true potential of the opportunity. The opportunity center is responsible for stimulating and activating the innovation activities for all the organization's employees. This is accomplished by providing training on problem-solving, creativity, and innovation methodologies to the organization. The opportunity center personnel provide one-on-one and group mentoring with the objectives of helping employees clarify and develop their ideas. They provide guidance and help the individual and organization to transform these concepts into tangible results.

In the development of a traditional value proposition, one generally understands that performance issues have been identified and are driving the development of the proposition. Most organizations have some limited ability to develop a value proposition, even if they do not call it such. Without this ability, many businesses would not be able to remain in business, since the nature of business requires it to be able to adapt to a changing market place. Some are better than others. The difference between an ordinary organization and an excellent organization isn't necessarily a question of the individual organization's ability to adapt, but to do so at a faster pace than the competition. Creative ideas/items that provide value-added products, services, and processes to the stakeholders can originate from any part of the organization, not just from a few functions like research and development, marketing, and industrial engineering. Among today's most popular approaches used to identify creative ideas and items is a methodology called "Benchmarking," which includes reverse engineering. Due to its popularity and proven usefulness over the past 20 years, we will highlight it as a source of creative ideas/items throughout parts of this book. But it is important to note that benchmarking results in less than 20% of the value propositions that are developed in a typical organization. For this reason, it is extremely important that today's organizations' value proposition process takes advantage of all potential sources that can generate creative ideas/items.

Analysis of these issues utilizing comparisons to leading practices and benchmarking as well as performance techniques are generally required upon completion of the leading practices and benchmarking, including gap analysis, identifying opportunities, developing key questions, and assembling data packages to assist in answering them. Finally, identification of potential answers for each opportunity should be developed. These discoveries eventually lead to a value proposition that is meaningful and useful. In some organizations, this process is left to a select group of employees, generally an engineering department or an engineering group. These days, however, we see more and more organizations utilizing the value proposition mechanism to allow innovators and "intrapreneurs" to hash out and refine innovative ideas that can help improve the bottom line of the company while creating a culture of innovation.

To discuss this in the context of the value proposition activities, the input to Activity 1 is always the recognition of an opportunity or problem that is currently being unmet, whether that is from ignorance or inability. Ignorance is, as Donald Rumsfeld so famously elaborated, "unknown-unknowns," while inability is more likely a "known-unknown," with unknown referring to "not knowing" how to improve a problem that we do recognize.

The value proposition approach, coupled with the opportunity center, not only allows us to develop answers to business performance issues but also allows the organization to consider new and radical approaches to problems in a structured format that allows all members of the organization, at all levels, to contribute expertise in areas in which they have expertise. They are the starting assumptions about future research results, and if ultimately proven true, then the potential answers become conclusions that should fully address each issue. These answers can vary from highly formalized problems identified by executives in the organization to the least senior rank and file member of the organization, who can bring a fresh set of eyes to long-standing problems.

The process includes the honest and complete identification of issues with the identification of opportunities to close performance gaps. Of course, this can work both ways, because opportunities can be identified before issues. As an example, one might identify a new software package that addresses an issue that the organization does not even know it has until the software is discovered. This is one of the valuable byproducts of the innovation age we are living in. No longer is it necessary for issues to be identified before opportunities (a command and control approach to management). Now opportunities can drive issues. Therefore, in this book, one should read each section as a "preferred" approach, but not necessarily be wedded to the order. In fact, a fundamental shift has taken place in business today; no longer is it necessary to say that solutions derive from

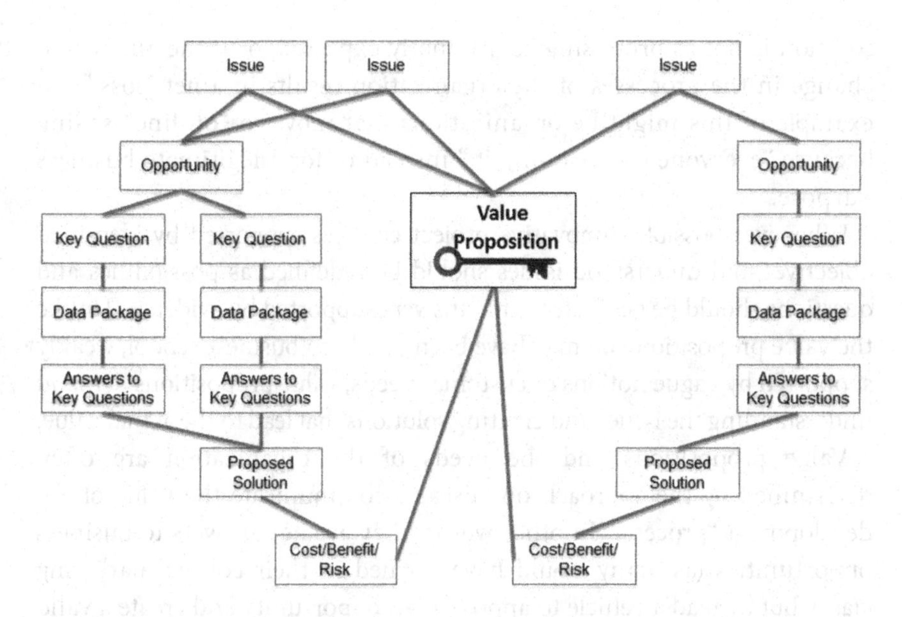

FIGURE 6.7
Alternative View of Value Proposition

problems. Instead, opportunities, in many cases, can drive issues. Or, with the advent of the Internet and low-cost software applications, solutions can affect a problem in a way that has never been possible before.

In the example of Figure 6.7, we might decide that "a reduction of customer losses that could generate $10 million per year in additional profit" would be our objective. This identification of an objective would then allow us to identify several issues that directly or indirectly affect the objective. As an example, we might be able to identify several ways in which customers lose money that we can actually do something about. Each issue is then turned and manipulated in a way that we will be able to define them as opportunities to achieve the initial objective. This leads to key questions related to the opportunities we have identified. Since questions demand answers, we could then take the key questions, develop support for the answers (through data and other supporting material), and develop the value proposition. Therefore, we must remember that depending on the organization, the steps may be sequential or modular. This depends a lot on the type of business you are in and the culture and the overall philosophy of your management. One is not necessarily better, and care should be taken when utilizing either approach, since a solution driving a process may be a very bad decision, especially if the solution is not as promising as originally expected, or if the amount of change in the processes of the organization results in a net "loss." An example of this might be organizations that move to "on-line" selling because "everyone else is doing it," instead of for a legitimate business purpose.

Whenever possible, innovative project changes supported by identified objectives and understood issues should be redefined as possibilities and questions should be generated with answers supported by evidence. Unlike the value proposition you may have been taught in business school, weakly supported by vague notions of customer needs, value propositions demand understanding the issues and creating solutions that lead to the actual value.

Value propositions and the needs of the organization are often determined by the approach one uses to communicate the value of the development "process." In other words, they are not answers to business opportunities (as many would have learned in their college marketing class), but instead a vehicle to approach an opportunity and create a valid business solution. One must also keep in mind that value propositions can be invalidated at any time. Therefore, when working through the value proposition methodology, a team or individual should pause at many points of the journey just to evaluate information as it evolves.

Process Grouping 3: Value Proposition and Tollgate II Activity Block Diagram

Typical activities that are included in Value Proposition are defined in the following Activity Block Diagram (see Figure 6.8).

3.1. Establish Opportunity Development Team meeting ground rules.

3.2. Review team charter, operating guidelines, and mission statement update if necessary.

3.3. Review requirements for value proposition's analysis.

3.4. Have members from the Opportunity Development Team explain their suggested changes and what impact they would have on value added from the stakeholder's standpoint.

3.5. Make a list of the positive and negative impacts the proposed change could have on primary stakeholder (customers, investors, and employees).

3.6. Review measurement data related to current state conditions and collect additional data to get required accuracy.

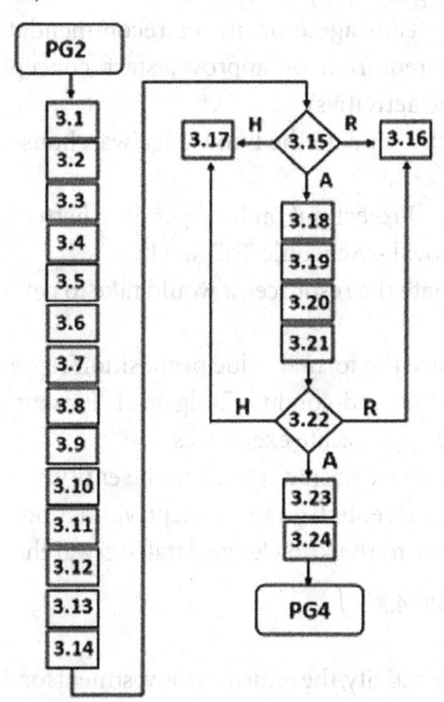

FIGURE 6.8
Process Grouping 3. Value Proposition and Tollgate II Activity Block Diagram

3.7. Make estimates on how current state will change over the next 12 months if the proposed changes are not implemented.

3.8. Using each individual proposed changes, estimate their measurable and psychological impact upon the organization and its stakeholders.

3.9. Estimate the measurable and psychological impact when all the changes are combined together.

3.10. Validate the implementation resource estimates provided by the Opportunity Development Team and adjust as necessary.

3.11. Have finance input the formal financial estimates and have marketing input the formal customer impact and sales price.

3.12. Get team and support areas to agree on the value-added content of the proposed changes and their implementation cost, quality, cycle time, etc.

3.13. Make a complete list of any special resources requirements.

3.14. Do a risk analysis and establish mitigation plans for medium level risk, including all safety-related risk.

3.15. Have team agree on major recommendations (reject, additional information required; or approve, start concept validation; or reject, terminate the activities).

3.16. Reject – Update the knowledge warehouse and terminate project/program.

3.17. Hold – Project put on hold pending better justification.

3.18. Approval – Schedule Tollgate II.

3.19. Estimate the resources it would take to get from Tollgate II through Tollgate III.

3.20. Prepare the formal value proposition report.

3.21. Schedule and conduct Tollgate II. Present form value proposition report to the appropriate executives.

3.22. Executive decision related to the entity.

3.23. Entity accepted, go to Concept Validation.

3.24. Update to the knowledge database/warehouse.

Forward to PG4.

During this activity, the return-on-investment for the high-priority changes will be calculated. It is important that both the positive and negative impacts that the individual change would have on the organization are defined and analyzed. Some companies pride themselves on being stakeholder centric. These companies take into consideration both the positive and negative impacts of all the stakeholders. Others don't consider human resource savings unless the resource is limited or assigned to another job that impacts the bottom line. In these cases, the savings from the human resource improvement is measured by the change in the bottom line as a result of their assignment or termination. Take time to develop, refine, implement, maintain, and then compare this to the value-added content that the changes will bring about. Based upon this analysis, the changes that have the biggest impacts, both direct and indirect, on the organization will be prioritized.

> *One of the most difficult problems organizations face is developing a comprehensive structure for measuring value added when they consider all the stakeholders.*
>
> **—H. James Harrington**

I cannot overestimate the difficulties in determining real value added. When you talk about hours saved, it is relatively easy to translate it into dollars. You just have to take into account that the money being paid to the employee falls into three categories: fixed overhead, variables overhead, salary, and benefits. In calculating savings per hour, there are two classifications of overhead: they are fixed and variables overhead. Fixed overhead is that need to be distributed across present product lines for support functions like personnel and finance. The fixed overhead that is billed against the hour is not eliminated when hours are reduced, as fixed overhead dollars are just redistributed to other activities. The variables overhead cost is legitimate savings. The benefits costs are sometimes almost as much as the salary paid to the individual. Most of the benefits costs are directly reduced when an hour is saved. What if you save an employee ½ hour per day but don't give him or her an additional assignment to do? Then there are no savings. It just provides them with a little more time at the coffee machine, play with their cell phones, or rest between jobs. Employee hours are saved only when productivity goes up or workforce is reduced. We have seen value propositions that claim there would be 100 hours saved a week, but there was no decrease in the cost to produce the output. A little savings spread over a lot of individuals typically results in little or no savings. To claim savings from work hours eliminated, there has to be both an increase in productivity and a decrease in processing costs.

Reducing inventory is worth the interest you pay on the value of the inventory reduced. But what if the savings is an indirect saving? How about things like cycle time reduction? If cycle time reduction doesn't decrease inventory or increase sales, what value is it? If increased customer satisfaction doesn't increase sales, what value is it? If improved morale doesn't increase productivity and decrease scrap and rework, what value is it? I have heard many CEOs say that they have implemented an improvement initiative that recorded big savings, but the results never got to the bottom line. Savings only occur when it's reflected in the organization's bottom line. Stopping doing one job to do another job that didn't need to be done before results in no savings. We recommend reading *Maximizing Value Propositions to Increase Project Success Rate* by H. James Harrington and Brett Trusko (published by CRC Press, 2014).

TOLLGATE II: CONCEPT APPROVAL

At this point in the cycle, a number of action plans have been developed and the team's best option has been selected. Estimations are now based upon hard data and prepared by knowledgeable individuals related to the proposed change activities. Improvement opportunities that will require a considerable investment or high risk and/or are critical to the future success of the organization should be presented to the executive team. The other initiatives will be presented to the impacted managers. Those that recommended budget changes should also be presented to and approved by the chief financial officer.

This review focuses on the detail contained in the value proposition. Particularly attention will be given to leading-edge innovative stations and technology advancements. Key factors to be considered are:

- Customer satisfaction
- Safety requirements
- Return on investment
- Projected technology advancements
- Improvements in output per employee
- Reduced cycle time
- Projected market share
- Improved functional performance
- Improved quality
- Potential new patent and patent infringement
- Competitive advantage
- Levels of risk
- Marketing and sales strategies
- Return on investment
- Long-range viability
- Last but not least, profit. For organizations licensed as profit-making organizations, the major reason they are in business is to make a profit. That is why they are classified as profit-making organization. If they don't, they should be treated as a hobby or a not-for-profit making organization

It's important to note that at this point in the PIC, the projected improvements are primary judgment calls, and there is little information related to the cost of change implementation. Although a number of theoretical approaches have been defined related to taking advantage of the opportunity, none of them has had their concepts validated. As a result, there still is a high risk of the recommended change failing to meet expectations and/or requirements. Projects that successfully pass this Tollgate are usually ones that the executive team would like to add to the organization's portfolio of active projects.

Frequently at the end of Tollgate II. Concept Approval, a project manager is assigned to follow the program through Business Case Analysis. At this point in time, the project is usually not part of the organization's portfolio of active projects, but it is a strong candidate to be included in the portfolio in the near future. As high as 60% of projects that enter Tollgate II are rejected and terminated even though they may add positive value to the organization.

PROCESS GROUPING 4: CONCEPT VALIDATION

By now, you have identified an improvement opportunity, defined ways to take advantage of this opportunity, and evaluated the potential solution to determine if it is value added to the stakeholders. Now two big questions – "Will it work? Will the solution bring about the potential savings as defined in the value proposition?" Failure at this point in the PIC results in considerable loss of money, time, and human resources. It usually means that the potential project is dropped or at a minimum redirected back to Process Grouping 2. Opportunity Development. This phase covers all the activities required to recognize potential improvement opportunities/problems, to creating a potential solution, and validating that the potential solution that will address the opportunities/problems.

Process Grouping 4: Concept Validation Activity Block Diagram

Typical activities that are included in Concept Validation are defined in the following activities (see Figure 6.9).

4.1. Review team charter and operating guidelines.

4.2. Review the requirements for Tollgate III and for Business Case Analysis procedures.

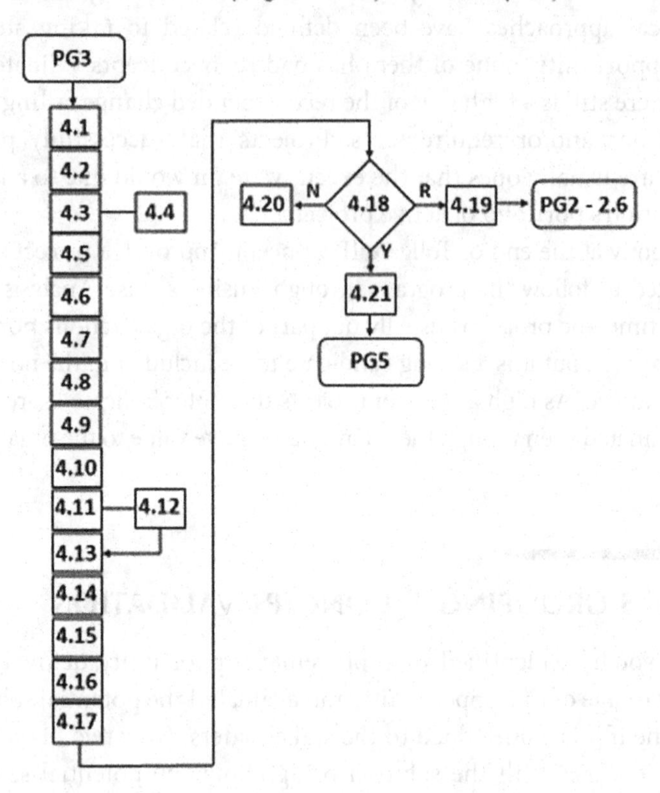

FIGURE 6.9
Process Grouping 4. Concept Validation Activity Block Diagram

4.3. Determine if additional functions need to be represented to the team to complete Tollgate III.

4.4. Add resources as required with management approval.

4.5. Make a list of data that is required for the Business Case Analysis and for Tollgate III.

4.6. Analyze all the engineering knowledge base related to the mission.

4.7. Develop and initiate your organizational change management activities.

4.8. Define model configuration for each desired improvement measurement or potential impact to other areas.

4.9. Prepare test plan and prepare software testing programs and update the project management plan.

4.10. Define the number of test samples that need to be evaluated in order to have sufficient confidence in the performance, safety, and reliability of the components when used in the specific application.

4.11. Order samples of all new technology components, and test key components for performance, function, and reliability.

4.12. Purchasing to order parts and equipment.

4.13. Construct models and laboratory test setups.

4.14. Have an independent observer validate the findings in the test method used to do the evaluation. (Typically, this verification would be done by quality assurance, or product engineering, manufacturing engineering, field maintenance, etc.)

4.15. Determine if the product will be subjected to any unusual conditions and measure performance. Example: Bombarded by radio frequency, subjected to high or low temperature, subjected to unusual vibration or high humidity environment.

4.16. Prepare a minimum of one pilot model where all of the interrelated performance parameters can be measured in relationship to each other.

4.17. Conduct experiments and test runs. If it is an upgrade of a current entity, run a control sample of the current entity in parallel with the experiments and tests so that data can be compared and improvements measured.

4.18. Did the innovative entity meet the goals and objectives defined for it in the project plan?

4.19. Reject – The concept needs to be refined, go back to PG 2.6.

4.20. No – Update database and terminate the project.

4.21. Yes – Update the knowledge warehouse.

PG5 – forward to Process Grouping 5 Activity Block Diagram.

During this activity, the proposed change is modeled, allowing new performance data to be collected. Modeling can be accomplished by building an engineering model of the change and submitting it to a number of conditions (e.g., temperature, humidity, vibration, and electronic external interference). The results can be used to project performance, failure rates, or reliability and customer satisfaction. Simulation models, both real and virtual, are also frequently used to validate the engineering and financial estimates.

Frequently innovative changes to upgrade software or processes will be modeled by running parallel activities for a short period of time.

This allows the results to be measured and compared. We are aware of situations where the additional attention given to a modeling activity has increased the efficiency and effectiveness of a process to the point that it outperformed the proposed innovative process. Without running a control sample, the proposed innovative process often is mistakenly implemented, thinking it would be an improvement when it was really not as good. Typical process innovations are measured in cycle time, processing time, accuracy, reduced costs, adaptability, customer satisfaction improvements, and space reduction.

SUMMARY OF PHASE I: CREATION

We have now completed discussing Phase I – Creation. By now the project team should have two or three potential solutions that need to be separated and evaluated to determine which has the best value added for the organization's stakeholders. It is important for you to understand that this is not the closeout of the processes included in Process Groupings 1 to 5, as there often is a need to go back to collect additional information or to repeat evaluations or to understand previous evaluations as the IPT goes forward into Phases II and III.

7

Phase II: Preparation and Producing

INTRODUCTION

With Phase I. Creation successfully completed, Phase II. Preparation and Producing begins. At this stage, the initiative becomes an approved project for the organization. Large complex and mission-critical projects are assigned to a project manager and become part of the organization's active project portfolio. The hundreds of minor improvement activities usually referred to as "continuous improvement" projects do not have the luxury of being managed by a professionally trained and experienced project manager. Not having well-trained, experienced project knowledgeable individuals in charge of these less demanding projects may be a major error in the present way we manage our continuous improvement and minor innovations within most of the organizations. Assigning professional project managers to more of the continuous improvement and minor innovative changes could have a significant impact upon reducing the cycle time, missed schedules, and cost overruns, while having a significant impact increasing the return on investment.

Phase II. Preparation and Producing consists of the following four Process Groupings and Tollgate III and Tollgate IV:

- Tollgate III: Project Approval
- Process Grouping 5: Business Case Analysis
- Process Grouping 6: Resource Management
- Process Grouping 7: Documentation
- Process Grouping 8: Production
- Tollgate IV: Customer Ship Approval

DOI: 10.4324/b22993-7

As the names imply, these four Process Groupings are very much standard activities. Only a few of them require a great deal of imagination or creativity. Most of them are Type 1 – Obvious Improvements or Type 2 – Minor Improvements. This doesn't mean that there isn't a continuous flow of improvement concepts being developed and implemented in these four Process Groupings. Usually, these changes are far less visible than the activities that take place during the four Process Groupings in Phase 1. Creation. Typically, the projects that are considered major in nature and receive high levels of attention are major upgrades to present products, process reengineering projects, or new software packages that are being applied to one or more of the four Process Groupings.

Many people feel that the creative/innovative cycle is over when they complete Phase I. Creation. This is far from the truth, as less than 6% of the total PIC cost is normally expended during all of Phase I. All of Phases I and II are an investment in the future that may or may not pay off. Phase I should have put us in a position where we have a high degree of confidence that continuing the project into Phase II will result in real value added to the organization, its customers, and their stakeholders.

The activities that typically occur in the four Process Groupings in Phase II. Preparation and Producing are:

- Tollgate III. Project Approval

  This is an analysis of the status of an individual potential innovative opportunity to determine if it meets all the requirements necessary to implement as part of the organization's activities. The proposed change has already been modeled and proven to create the desired result. Risks have been defined and mitigation plans have been prepared for high and medium-risk exposures. All the resources required have been accurately defined, and both the positive and negative impacts of the activity have been identified and are acceptable. The only thing that is keeping the project/program from being activated, funded, and included in the strategic plan is an evaluation to determine if this is the best way to invest in the organization's limited discretionary spending. This decision will be made during Process Grouping 5. Business Case Analysis.
- Process Grouping 5. Business Case Analysis

  This is where you get approval, financing, budget, performance specifications, human resources, schedules, and executive support to an individual project/concept. It is usually a go or no-go decision activity.

- Process Grouping 6. Resource Management

  This is where you transform a budget into money, people, facilities, and materials required to develop the concept so that it can be produced in the required quantities. This is typically the point where an official project manager is assigned and additional staff is added to the IPT.
- Process Grouping 7. Documentation

  This is where the rough notes from the engineering notebook are transformed into engineering specifications, which are then released as product specifications and requirements. These product specs are then used to document the processes and procedures that will be used to produce the output and control its efficiency and effectiveness.
- Process Grouping 8. Production

  This is where the manufacturing documentation (routings, training procedures, operating instructions, test procedures) becomes the basis of decisions to move on to the next activities, such as approving to build or subcontract, what equipment and data collecting systems are installed, and what facilities to set up. It also includes the training of the people who will produce the output and the suppliers who provide input to the process so that the process cost is minimized and the external consumers receive output that meets and preferably exceeds their requirements and at a price they consider reasonable. The objective is to maximize the value added for all of the stakeholders.

  During this phase, the proposed changes are analyzed to determine if they should be included as part of the organization's portfolio of active projects. Once the change becomes part of the organization's portfolio of projects, resources are set aside to support the change process, to create the necessary engineering and manufacturing documentation, to validate the acceptability of the production outputs through a series of manufacturing process model evaluations, and to start shipping to an external customer/consumer.
- Tollgate IV. Customer Ship Approval

TOLLGATE III: PROJECT APPROVAL

Tollgate III is closely tied into Process Grouping 5. Business Case Analysis. The proposed changes have completed the concept evaluation activities and are getting ready to prepare a detailed Business Case Analysis if the project successfully meets the requirements defined in Tollgate III. Business Case Analysis normally differs greatly from a value proposition that was prepared during Phase I. Value propositions primarily focus on the value of the project. Business Case Analysis focuses on how the project fits into the priorities and opportunities of the organization. It takes into consideration if the proposed project estimates are based upon sound statistically available information and if the cost estimates had been generated by the financial department.

Officially completing Tollgate III indicates that the program/project meets the requirements and conditions defined in the project mission statement, project plan, and the related entity's documentation. It also indicates that the project is recommended and should be included for consideration in the organization's portfolio of active projects/programs. Completing Tollgate III does not mean that the organization should use its limited resources so that the program/project would be included in the present portfolio of active projects. Frequently a project that successfully completes Tollgate III will be rejected during the Business Case Analysis or be put on hold until resources become available as other higher priority projects are completed. For more information related to portfolio management, we recommend reading the book entitled *Effective Portfolio Management Systems* (CRC Press 2016).

This Tollgate assessment penetrates deeply into the business aspects of the project and how it will impact return on assets, stock prices, and market share. Much of the effort is focused on how the change will impact current and future outputs and the organization's reputation.

Tollgate III: Activity Block Diagram

The following is a list of typical activities that take place during Tollgate III. We will first focus on Tollgate III, which is designed to focus on evaluating the proposed project/program effect on the organization's stakeholders. We will then focus upon performing a Business Case Analysis in the next section of this book. The Business Case Analysis objective is to control the use of the organization's resources in order to produce the most value-added content as viewed by the organization.

Frequently projects and programs that pass Tollgate III will be terminated during the Business Case Analysis because the particular project or program does not make the best possible use of the organization's resources or its strategies. Usually, the makeup of these two executive teams is very different. In the case of Tollgate III, the executive team is usually made up of executives who are impacted by the project/program. The executive team that evaluates and makes the decisions related to Business Case Analysis usually consists of representatives from each of the functional units that make up the executive staff because it impacts present, proposed, and future activities in all areas of the organization.

This Activity Block Diagram will only cover the activities going on in support of Tollgate III (see Figure 7.1).

5.1. Form Tollgate III Team.

5.2. Appoint Project Manager – Team leader should be assigned to pull together the Tollgate III analysis to be presented to the appropriate

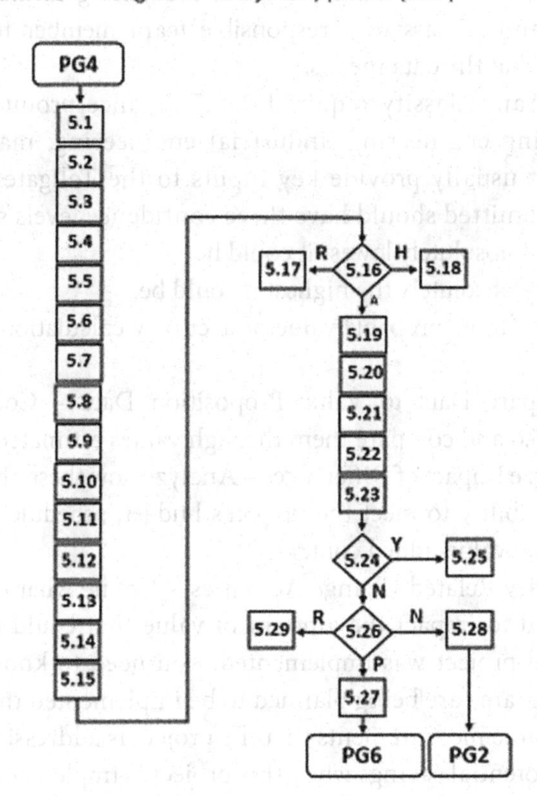

FIGURE 7.1
Phase II. Preparation and Producing and Tollgate III Activity Block Diagram

executive team for approval. Sometimes this is a project manager or strategic planning manager.

5.3. Prepare Team Charter – Review the creation team charter, value proposition, and requirements for Tollgate III. Plus review the knowledge management file related to the activity available to the team members.

5.4. Finalize Project Objectives – Review and discuss the product's specifications and performance objectives.

5.5. Meeting with Creation Team – Have each creation team member discuss what additional information they had collected since inputting into the value proposition.

5.6. List Key Estimated Values – Make a list of the key estimated values used in the value proposition and record its values.

5.7. List Required Information – Make a list of the information required to complete Tollgate III.

5.8. Team Members Assignments – For everything on the list prepared in 5.5, 5.6, and 5.7, assign a responsible team member to collect and understand what the data means.

5.9. Collect and Classify Required Data – Finance, production control, manufacturing engineering, industrial engineering, marketing, and procurement usually provide key inputs to the Tollgate III. All data estimates submitted should have three confidence levels stated.

- Level 1 – absolutely lowest it could be.
- Level 2 – absolutely the highest it could be.
- Level 3 – most probable value or accuracy calculations (accurate to ±10%).

5.10. Compare Data to Value Proposition Data – Compare values collected in 5.6 and compare them through values estimated in 5.9.

5.11. Analyze Impact of Differences – Analyze how these changes impact the project's ability to meet the project's budget, schedule, performance objectives, and value-added content.

5.12. Identify Related Change Activities – Define what other changes are scheduled to impact the amount of value that could be realized if the proposed project was implemented. You need to know what other projects/programs are being planned to be implemented that will or can impact the same measurements that the project is addressing that would reduce the potential savings when this project is implemented. Example:

It takes 12 employee hours to produce an output. The project being evaluated is going to apply lean concepts to the process reducing the time to produce the output to eight employee hours. This is a savings of four employee hours. Manufacturing engineering already has approved a project to automate the same process reducing the employee hours from 12 to 6 for a savings of six employee hours. If both projects were implemented, the organization will be expecting a value added of 10. (Lean saves 4 hours + the automation saves 6 hours = 10 employee hours saved total.) These are not the results that the organization would actually benefit from. If Lean was installed first, automation could probably only save two hours rather than six hours and could not be justified. The interaction between projects going on within the organization is an extremely important part of Business Case Analysis activity, or the results will not filter through to the bottom line.

5.13. Analyze the impact of 5.12 to determine how data collected in 5.12 will impact the data collected in 5.11.

5.14. Calculate Value Content – Calculate the value content of the proposed change as it is viewed from each of the stakeholder's and the organization's viewpoints.

5.15. Estimate Key Value Added – Use the information collected to re-estimate the values for the key data required to decide to continue or drop innovative opportunities.

5.16. Define Team Recommendations. Tollgate III team to agree on the recommendation.

5.17. Reject, Terminate Project – Update the knowledge warehouse and terminate the project.

5.18. Hold – Need additional data, return to Process Grouping 3.

5.19. Accepted, Outline Final Report – Prepare outline for Phase III final report.

5.20. Prepare Final Report – Assign related team members to prepare their part of the final report.

5.21. Executive Sponsor Report Review – Schedule a meeting before the formal Tollgate III review with the executive sponsor and present the findings to him so he can sign off on the formal report.

5.22. Schedule Tollgate III Part 1 – Schedule Tollgate III Part 1 review meeting. Invite all impacted managers and key personnel. Plan a one-hour meeting focusing on minor impacts to the bottom line.

5.23. Conduct Tollgate III– Conduct the Tollgate III executive meeting emphasizing project's potential value added both real and intangible. Also focus on your estimate for the use of additional resources prior to start of shipping to customers. In deciding which projects/programs will be added to the organization's portfolio, I like to use the absolutely lowest possible value-added minus the absolutely highest possible project-related costs to calculate return on investment. We use this approach because most companies' cost estimates are lower than actual, and value-added estimates are higher than actual.

5.24. Was the project/program terminated?

5.25. Yes – Update the knowledge warehouse and terminate the project.

5.26. Results of Tollgate III Part 1 – What was the executive team's decision?

5.27. Passed, but put on hold for a final decision in Business Case Analysis. A decision cannot be made on a potential product until available resources have been established.

5.28. Hold – More data needed. Additional information required for approval. Return to Process Grouping 2. Opportunity Development.

5.29. Rejected Project – To be shut down after information warehouse is updated.

Go to PG5-5.40.

Tollgate III: Project Approval Supports the Business Case Analysis

Tollgate III and Business Case Analysis are two separate and very different activities.

- Tollgate III. Project Approval focuses on evaluating if the project met the requirements defined in its mission statement and project plan. It is a very detailed focus on how the project/program will benefit the stakeholders and the probability of it being successful. This should be a detailed focus review on the potentially innovative project/program, including an analysis of how the project will impact value added for all the stakeholders. Successfully completing Tollgate III makes the project eligible to be included in the Business Case Analysis review of all the proposed potentially innovative projects/programs and current activities. Business Case Analysis usually takes place during a different meeting than the one held for Tollgate III.

- Business Case Analysis meetings discuss the merits of investing resources in the present and proposed entities. This is usually a high-level executive meeting, as it covers a broad range of presently active or approved entities and all of the projects /programs that have successively completed Tollgate III. It is the activity where current resource allocation is assigned to proposed potentially innovative projects. Based upon past experience, there are often a number of proposed projects that could produce real value-added results but are not approved due to resource limitations or resource assignment to other projects that have a bigger impact upon the organization's present and future status.

Comparison will be made between the proposed change and the resources that would be expanded in current products and previously approved projects. Even currently approved entities that are being shipped to customers may be drastically cut back if applying the resources plan for the current entity is better utilized by applying it to the potential innovative project. This is a no-holds-barred meeting because those projects that successfully complete the Business Case Analysis will drive the future of the organization as well as maintain the current revenue stream. Great care needs to be taken related to the product selection and the announcement timing to keep ahead of the competition or in many cases just to maintain the status quo.

PROCESS GROUPING 5: BUSINESS CASE ANALYSIS

This is where you get approval for financing, budget, equipment, performance specifications, human resources, schedules, and executive support to an individual project/concept. It is the point where the project becomes an official part of the organization's portfolio. It is usually a go or no-go decision activity.

Usually, there are a series of meetings held where the Business Case Analysis is reviewed for each proposed project/program. There also is a review of the status and value-added content for each of the major active projects/programs. Then a meeting of the executive team reviews proposed new and present active projects/programs to determine the best use of the organization's resources. This meeting can result in present projects proposed being terminated or being funded and added to the list of active projects/programs in the organization's portfolio.

Inputs to Process Grouping 5: Business Case Analysis

- Input(s) Process Group 5:
  - Portfolio development leader
  - Portfolio development team members (if appropriate)
  - The business cases (proposed projects) to be analyzed
  - Related information from the knowledge warehouse
  - Project/program charter, goals, and objectives
  - Results from concept validation and Tollgate III
- Activities in Process Group 5:
  - Business case validation
  - Validate accuracy of projections
  - Document performance and project resource requirements for each project/program
  - Analysis of business cases that do not require additional resources
  - Select and use a set of criteria aligned to the organization's mission, strategic plan, and long-range objectives to classify and rank the business cases
  - Defined high- and major-risk areas and associated mitigation plans
  - Determine which classification model to use – Qualitative, Quantitative, or Blended
  - Review of present portfolio projects and programs
- Output(s) from Process Group 5:
  - An executive committee review and approved ranked-ordered list of projects and programs based upon their potential added value/impact
  - The organization's portfolio is adjusted
  - Budgets are adjusted to reflect changes in portfolio and current activities

As part of a typical Business Case Analysis cycle, each function should have submitted a set of business cases that they would like to start during the next business cycle. (Note: For details, see the two books *Value Proposition Development* and *Business Case Development*, published by CRC Press.) On some occasions, the functional units submit projects/program business cases for inclusion in the active approved activities within the organization between budget cycles. In these cases, the Portfolio Development Team handles them as a special case. These situations are discouraged, but in today's organization with a very fast changing environment, it is practically impossible to eliminate these special evaluation cases and still have the organization function effectively. We try to avoid as many of these cases as possible as they can become very time-consuming and costly. Typical ones that are considered are a reaction to competition's unsuspected release of the new advanced innovative product that obsoletes the organization's outputs.

Based upon our personal experiences during a budget cycle, a number of improvement opportunities are identified that have *not* gone through the Value Proposition Development stage or the Business Case Development stage either. Often it is *not* practical to ignore these improvement opportunities, and as a result, the portfolio development leader will need to work with the individual functional area that is recommending or "nominating" these improvement opportunities to, at a minimum, prepare the data that is required for a business case so that the improvement opportunity can be fairly considered along with the other business cases. Often these last-minute improvement opportunities actually turn out to be "pet projects" sponsored by key executives within the organization, and ignoring these key inputs could be politically "sensitive" and detract from the organization's potential performance. Unfortunately, these improvement opportunities have a tendency to increase the length of the budgeting cycle and, as a result, should be discouraged whenever possible. In any case, no project/program that impacts the budgeting cycle should be considered unless it has an executive sponsor, a projects/program leader (champion), and a sound resource requirement and an estimated value-added analysis complete and submitted with the request.

The portfolio development leader and his/her portfolio development team will focus their attention on classifying and ranking the proposed projects/programs business cases to develop the prioritized list of potential projects that will be considered to make up the approved portfolio.

Business Case Validation

During Tollgate III, the IPT should review each proposed project/program to ensure its business case is well-developed and includes practical and realistic estimates related to its goals, performance objectives, timing, and resource requirements. At a very minimum realistic goals, performance objectives, timing, and resource requirements must be documented or the project/program should not be considered by the Executive Team for being included as an active project within the organization.

Document Performance and Project Resource Requirements for Each Project/Program

We find that while the Portfolio Development Team is reviewing the Tollgate III, individual project and programs business cases are an excellent time to prepare a list of all the projects being evaluated and record what the projected impact on the organization's performance and the projected resource consumption. We also suggest you record estimated implementation time and any risks that the group that prepared the business case defined as impacting the project/program. This provides an effective bird's eye view of all of the proposed projects/programs being evaluated.

Business Cases That Do Not Require Additional Resources

Many of the business cases that are completed do *not* require additional resources because maintenance and problem resolution for current established processes is included in the functions annual budget. Many of the identified innovative opportunities can be implemented within the normal activities that go on within the function and already in the approved budget. (Example: Product engineering could have resources already budgeted to correct problems or make small changes to a current product. This would include activities like redesigning a part that is presently a steel machine part and replacing it with a plastic molded part.)

Another example would be when an operator suggested a change in the production routing that would result in reduced cycle time, setup time, and defect rates. These improvement efforts requiring resources to evaluate and implement are activities that are normally part of the day-to-day job responsibilities of the individual organization and automatically included in the annual budget. As a result, these business cases do *not* need to be considered as part of the organization's project portfolio. Only those business cases where the scope, magnitude, and impact fall outside of the normal job responsibilities or new projects/entities/programs that have successfully completed the Business Case Analysis should be considered for inclusion in the organization's portfolio of projects and programs. Usually, for these normal job responsibilities types projects, a value proposition at the most is all that needs to be prepared and approved by the function's management team. In many cases, approval of these types of activities are either automatically approved by a memo or a meeting where the activity is discussed and approved.

Criteria to Rank Business Cases

The executive team should select and use a set of criteria aligned to the organization's mission and strategic plans to classify and rank the business cases. The first step is to perform a general evaluation of each of the business cases to evaluate how it fits into the organization's overall business structure. While this step should have been done as part of preparing the proposed business case, we have found it helpful to double-check it at this point in the portfolio development stage of the lifecycle. Each of the business cases should be reviewed against the following nine items to determine if it is in minimum compliance with the related criteria (see Table 7.1).

TABLE 7.1

Business Case Compliance with Key Business Considerations

#	Item	Criteria
1	Mission statement	Must comply
2	Policy	Must comply
3	Vision statement	Should comply
4	Values	Must comply
5	Strategy	Must comply
6	Critical for success factors	Need not comply
7	Business objectives	Must comply
8	Organizational goals	Should comply
9	Strategic business plan	Must comply

Any business case that does not meet any one of the six "Must comply" criteria should be dropped from further consideration for the organizational portfolio. Any business case that meets all six of the "Must comply" criteria items but has an item that doesn't meet one of the two "Should comply" criteria should still be given additional consideration during the evaluation cycle.

In addition, every business case should have, at a minimum, a sustaining sponsor identified who will be held responsible for the success or failure of the project going forward.

- Determine which classification model to use – Qualitative, Quantitative, or Blended

  Now there is a more granular and detailed approach to arriving at the criteria needed to classify and rank the projects and programs, one of three detailed classification or selection models: "Qualitative," "Quantitative," or "Blended."
- A detailed selection model using "Qualitative" criteria provides an "anecdotal" or "subjective" perspective.
- A detailed selection model using "Quantitative" criteria provides an "empirical" or "objective" perspective.
- A detailed selection model using a "Blended" approach draws from both of the other two criteria providing a "dual" or "hybrid" perspective.

Depending on the set of criteria chosen to classify and rank the projects and programs as an Input, you should determine which one of three detailed classifications or selection models you're going to apply to arrive at that conclusion: a "Qualitative approach," a "Quantitative approach," or a "Blended approach."

(Source: *Effective Portfolio Management Systems*, Voehl, Harrington, and Ruggles, published by CRC Press, 2016)

Results of Business Case Approval

Passing the Business Case Analysis opens the floodgates as resources are poured quickly into the process to complete the massive amounts of documentation necessary and to establish a production system that will meet the massive amounts of purchases that will accompany first-customer ship to external customers and consumers. On top of that, with the very short-product cycle times, the first part has to be as good as the last part produced. You no longer have the luxury of shipping unknown products to your customers and then going back and repairing your design or your production facilities. For by the time you completed the corrective action cycle, somebody else's new product will have stolen the market. As much as 35% of the projects that are submitted to Tollgate III. Project Approval are rejected. Your organization's reputation is based upon the outputs you use to deliver more than the products you are delivering. The results of delivering unsatisfactory or marginal outputs today results in bankruptcy tomorrow.

Any product or program that successfully completes the Business Case Analysis and isn't a message success is a testimonial to a significant executive blunder.

—H. James Harrington

Your organization's reputation is based upon the outputs you use to deliver more than the products you are delivering. The results of delivering unsatisfactory or marginal outputs today results in bankruptcy tomorrow.

—H. James Harrington

Process Grouping 5: Business Case Analysis Activity Block Diagram

One of the key factors in deciding which projects become part of the organization's portfolio is the quantity/number of resources that are available or can be made available for current ongoing activities and the new projects/programs that will be added to the portfolio. Resources can be divided into following three distinct types of activities:

1. Resources used to produce and deliver current products and services to the external customer.
2. Resources used by support functions like HR, marketing, research and development, production control, manufacturing engineering, industrial engineering, information services, etc. These are primarily your fixed overhead resource users.

3. New not currently funded projects, products, services, public relations, potentially innovative programs, etc. These are activities that are primarily focused on improving the organization's present performance but not directly related to a specific output that is going to a customer/consumer.

During the Business Case Analysis, projects and programs are compared and prioritized in relationship to all other ways that the organization is considering using its resources. This is a very competitive situation where only the real business opportunities should survive. A small group of high-level staff members will be assigned to establish the amount of discretionary resources that are available for assignment. They will then analyze all of the requests for resources and make recommendations on how these resources should be utilized. The results of their findings will be presented to the executive team so as to make a final decision related to the use of discretionary and presently assigned resources. This is a critical point in the innovative cycle because often projects/programs that will provide positive value added to the organization may be dropped as a result of the Business Process Analysis in favor of other projects that have a better return on investment. The Business Case Analysis Activity is usually performed once or twice a year when a group of them are analyzed at the same time. This analysis provides the database that is used each time a proposed project is requesting additional resources.

The following is a list of typical activities that take place during the Business Case Analysis part of the budget cycle (see Figure 7.2 for Business Case Analysis – Activity Block Diagram).

5.40. Total resource list – Prepare a list of the resources and the related magnitude that are available to the organization. Typical resources: number of employees and their skills, human capital analysis, approved financing, equipment and its availability, available facilities, approved suppliers, patents, inventories, marketing channels, etc. (NOTE – This activity needs to be done only one time when a total new budget is developed for the organization each year. The data will remain essentially constant between budget cycles and will be used for all projects and programs that are subjected to Tollgate III during that time period.)

5.41. Current operations resource requirements – This includes employees, skill levels, dollars, facilities, suppliers, materials, etc. (Example: Resources

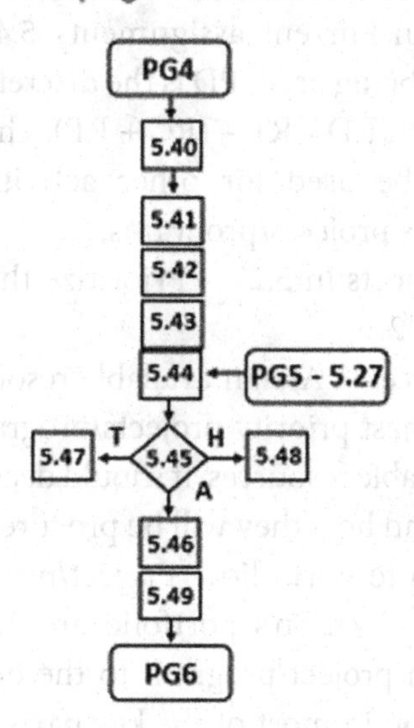

FIGURE 7.2
Process Grouping 5. Business Case Analysis Activity Block Diagram

assigned to HR, Quality Assurance, Production Control, Procurement, Research and Development, and certainly don't forget the resources required to continue to produce the output that you are presently delivering.) Other than the resources required to physically deliver present outputs to external customers, these and similar functions within your organization are the primary driver of your fixed overhead. In many organizations, today they consume more of the organization's resources than the resources that deliver the organizations outputs. (NOTE – This activity will need to be updated to reflect a change in the current portfolio. It will not be necessary to go back and pull together the total database.)

5.42. Present portfolio resource list – Make a list of all current projects/programs included in the portfolio of projects and the resources that they will require. (NOTE – This activity needs to be done only one time when a total new budget is developed for the organization each year. The data will remain essentially constant between budget cycles and will be used for all projects and programs that are subjected to Tollgate III during that time period.)

5.43. Calculate 5.40 (total available resources RT) minus 5.41 (current activities assigned RC) and 5.42 (resources assigned to active projects

RP) – Subtracting from the total available resources defined in 5.40 the resources assigned in current assignments 5.41 and 5.42, how much project resources are being used. PD is the discretionary resources that are available for relocation RD = RT – (RC + RP). These are the discretionary resources that can be used for other activities like supporting the potentially innovative projects/programs.

5.44. Prioritize projects in 5.27 – Prioritize the projects/programs that were put on hold in 5.27.

5.45. Assign resources – Assign available resources to the new projects starting with the highest priority projects/programs. When the executive team runs out of available resources, it should decide if additional resources should be acquired and how they will be procured.

5.46. Project added to portfolio – Project/program that is approved to be added to the organization's portfolio and to update the knowledge warehouse. Adding a project/program to the organization's portfolio of projects triggers action in most of the key parts of the organization. In a product-producing organization, it could trigger action in functions like the project management office, human resources, information services, manufacturing, manufacturing engineering, industrial engineering, product engineering, procurement, quality assurance, production control, after-sales services, sales, and marketing. We will not present an Activity Block Diagram for each of the functions, but every innovative or progressive organization should prepare them in keeping with the organization's unique operations. An understanding of the activities that take place during the development and implementing of an innovative entity and their interactions greatly strengthens the organization's Innovative Management System.

5.47. Project terminated – Project will be terminated after the knowledge management system is updated.

5.48. Project put on hold – Project put on hold waiting for more information and/or available resources.

5.49. Updated knowledge warehouse.

PG 6.0 – start Process Grouping 6. Resource Management.

Business Case Analysis is an evaluation of the potential impact a problem or opportunity has on the organization to determine if it is worthwhile investing the resources to correct the problem or take advantage of the opportunity. An example of the results of the Business Case Analysis of a software upgrade could be that it would improve the software performance as stated in the value proposition, but (A) requires 5% more tests processing time, (B) would decrease overall customer satisfaction by an estimated two percentage points, and (C) reduces system maintenance cost only $800 per year. As a result, the Business Case Analysis did not recommend including the project in the portfolio of active programs. Often the business case is prepared by an independent group, thereby giving a fresh unbiased analysis of the benefits and costs related to completing the project or program.

Summary of Process Grouping 5: Business Case Analysis and Tollgate III

During these two activities, you should now have proposed opportunities to determine how the organization's resources should best be utilized. Approved projects should have detailed project management packages prepared for them. Projects that successfully complete this analysis are usually funded through first-customer ship and become part of the organization's portfolio of active projects. To get a better understanding of the business plan analysis activity, we recommend reading *Effective Portfolio Management Systems*, published by CRC Press, 2015.

PROCESS GROUPING 6: RESOURCE MANAGEMENT

As we start Process Grouping 6. Resource Management, the executive team and the board of directors have agreed that a specific set of resources have been committed for the development and implementation of each of the approved project/program listed in the organization's portfolio of active projects. Typically, these resources are as listed here:

- Human resources
- Financial resources
- Floor space resources
- Equipment resources

These allocations do not mean that the resources and skills are now available to be assigned to these approved projects. It's like parents telling their daughter that she can go to the dance but no one has agreed to take her yet. It is now her responsibility to go out and scavenge an invitation from a boy who has a driver's license. This is where you transform a budget into money, people, facilities, and materials required to develop the concept so that it can be produced in the required quantities. Due to the unique processes needed to fill the requirements, we will present unique Activity Block Diagrams for each of the following types of resources:

- Human resources (staffing)
- Facilities resources
- Financial resources
- Equipment resources
- Floor space resources
- Facilities setup resources

This is typically the point where an official project manager is assigned and additional staff are added to the Innovative Project Team (IPT). During this activity, the resources that are required for the approved project are put in place.

In small and startup companies, financing usually becomes a major problem. Initially personal funding is used, then family funding, angel funding, and borrowing from banks are all legitimate sources. People resources also present a problem for both the small and large companies. Although there are sufficient people out of work today to fill all the available jobs, there's a big shortage in fields like product engineering, programming, and manufacturing engineering. Finding the right suppliers at the right price that can produce the correct item and do it on schedule in small lots is another problem that an organization faces during this activity. The last major item addressed in this activity is facilities. Not having the right equipment or the floor space required to support the output is a problem that must be addressed early in the product cycle.

Inputs to Process Grouping 6: Resource Management

The following is a list of typical inputs that are used in during Process Grouping 6. Resource Management.

- Output from Tollgate III
- Present budget

- Entity objectives
- Key measurements
- Personnel status
- Facility utilization analysis
- Financial reports
- Equipment utilization
- Certified supplier list
- Relevant information in the knowledge warehouse

Process Grouping 6: Human Resource Staffing Activity Block Diagram

Of all the problems facing a new innovative project the availability of qualified human resources is often the number one problem, followed very closely by financing, particularly in smaller organizations that do not have deep pockets. Let's look at the situation related to human resources. This can be a twofold problem:

1. You could have too much work that needs to be completed for the number of employees you have available to do the work.
2. You can have enough employees to do the work, but they do not have the skills required to do the work.

Now, at the completion of Tollgate III the organization has authorized the project team to use the resources they require to complete their assignments successfully. But–and that's a big "but"! – although the human resources have been approved, most of them are not presently assigned to complete the project. Experienced project managers will have developed their schedule taking into consideration the amount of time required to bring the required staff on board and get them trained. We've seen projects where the delay to require approved staff has cost a three-month project to slip six months. It is for this reason it is imperative that your staffing process flows need to be well documented, evaluated, and streamline. Minimum and maximum staffing time needs to be well understood by any individual who is managing an innovative entity during this development and implementation cycle.

Until you are fully staffed for the point you are in the project development and implementation plan, you are driving a 6-cylinder car that only has 4 cylinders working and the executive team cannot understand why you're not keeping up with the rest of the pack.

—H. James Harrington

The following is a list of typical activities that take place during Process Grouping 6. Human Resource (Staffing) Management cycle (see Figure 7.3).

6.1. Project manager assigned – A specific individual is assigned and is now managing the project.

6.2. Project kickoff meeting – The project manager will call a meeting where representatives of the organizations that will be impacted by the project should have at least one representative in attendance. The

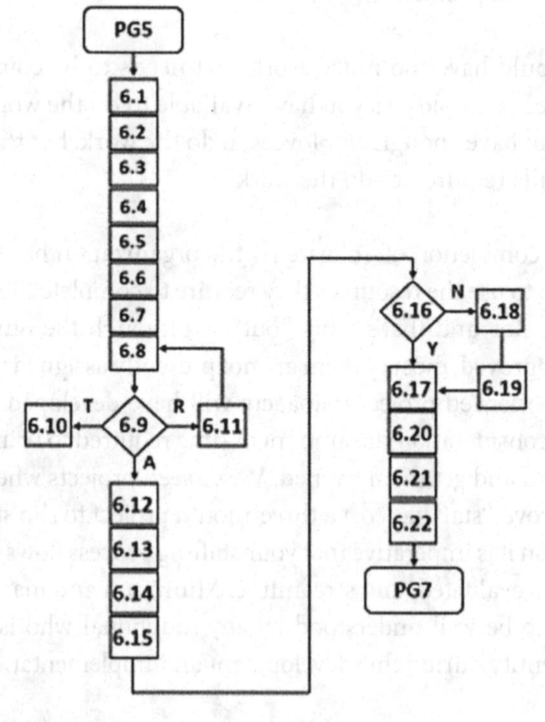

FIGURE 7.3
Process Grouping 6. Human Resource Activity Block Diagram

Tollgate III information will be presented and future activities will be discussed.

6.3. Assign team representatives – Representative from major involved functions (Example: product engineering, marketing production control, manufacturing engineering, and strategic planning).

6.4. Set key project dates – Set key start and end project dates required to meet project objectives, production quantities, estimates, and outsourcing plans.

6.5. Project review meetings – Sample schedule of project review meeting where the status and problems related to the project will be discussed and addressed. At a minimum it should be scheduled just before the completion of each process grouping. The frequency of the meetings is usually determined by the complexity of the project, the length of the project, and the number of different organizations that are involved in setting up the project.

6.6. Update the project team charter – The charter should define the organization's expectations, responsibilities, and accountabilities related to the specific project. It should be updated by the team and approved by the executive sponsor.

6.7. Prepare budgets – Prepare budget based upon Tollgate III requirements.

6.8. Prepare/update project plan and work breakdown structure – A detailed project plan could be prepared and approved by the executive sponsor. For major and complex projects, the work breakdown structure is one of the most important documents that the project team will generate because it shows all of the individual players, activities, outputs, and interdependencies.

6.9. Executive sponsors review decision – Results of executive sponsor review of project plan and work breakdown structure (WBS).

6.10. Project terminated – The executive sponsor could decide to update the knowledge warehouse and terminate the project.

6.11. Rejected, go back to 6.7 – Additional information or refinement is required before the project will advance.

6.12. Accepted – The executive sponsor approved the charter, project plan, and work breakdown structure.

6.13. Update computer system – Update computer system to reflect budgets, WBS, and project plan.

6.14. Acquire staffing per WBS.

6.15. Define staff requirements – Define the number and skills of permanent and temporary employees that are needed.

6.16. Availability of internal staff – Is a qualified internal staff available?

6.17. Assign individuals – Yes, place in the required departments/groups.

6.18. No – Hire external personnel.

6.19. Hire temporary and permanent staff as required.

6.20. Conduct new employee training.

6.21. Personnel resources fulfilled.

6.22. Update knowledge warehouse.

Output goes to Process Grouping 7. Documentation.

Process Grouping 6: Facilities Resource Management Activity Block Diagram

Facilities Resource Management includes providing the correct amount of space, utilities, layout, and equipment that the project requires making maximum effective use of the resources that the organization has already acquired. It has to include the space for both support and production facilities, including warehousing and storage. Facilities is often a very important consideration when it comes to this determining what will be produced within the organization versus subcontracted or procured. This is a resource that expands as approach is being developed and produced and then shrinks down as production falls off and new entities are phased-in. We have subdivided facilities resource management into the following:

- Floor space resource management
- Equipment resource management

Floor Space Resource Management Activity Block Diagram

The following is a list of typical activities that take place during Process Grouping 6. Activity Block Diagram for Floor Space Facility Resources. It lists the major activities that would typically go on in selecting space for a new project and setting up the facilities (see Figure 7.4).

6.30. Industrial engineering reviews space requirements defined in Business Case Analysis and updates as required.

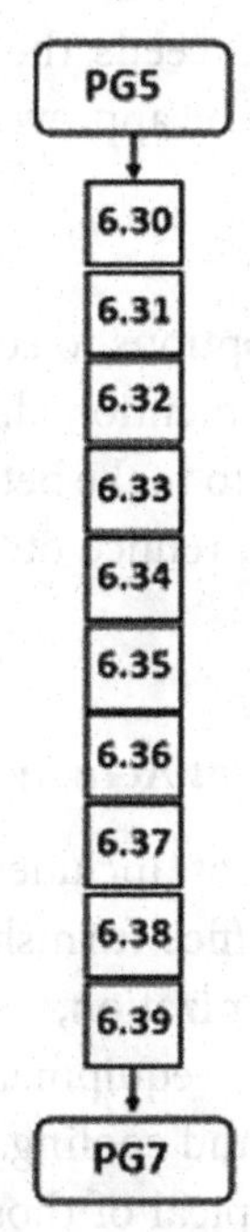

FIGURE 7.4
Process Grouping 6. Resource Management for Floor Space Activity Block Diagram

6.31. Determine if the product lends itself to rearranging the production area for each different style entity.

6.32. Define environmental conditions requirements (temperature, humidity, air quality, electrical and audio noise containment, security, etc.).

6.33. Review property layouts and limitations.

6.34. Review to determine availability of space, utilities, layout, and equipment.

6.35. For all special machines and equipment, get the machine diagram to determine that the power connections are properly mounted on the floor and that consideration be given for the conveyor equipment.

6.36. If space is not available, additional outsourcing should be considered.

6.37. If there is still a problem, determine if rental or construction is preferable.

6.38. Review alternatives with the executive sponsor, agree on an action plan, and update the knowledge warehouse.

6.39. If the action plan cost exceeds the approved budget process, a project variance request needs to be approved.

Output goes directly to PG7.

(Note: There are many other options to solving the facilities limitation problem than the ones previously mentioned. Some of the other approaches could be: working a second shift to make better use of space, lower output expectations, use of home office, reduce output quantity schedule, move inventory storage off-site, etc.)

Equipment Resource Management Activity Block Diagram

Equipment resource management includes the selection, installation, utilization, and disposal of all office furnishings and equipment, mobile equipment used by the organization, tools, machinery, computer equipment, and communications equipment. It does not include basic utilities like electricity, heating, and cooling.

The following activities are typical of those that take place in order to provide the proper equipment and utilities needed to support the approved project (see Figure 7.5).

6.40. Define new or additional equipment needs including automation and software requirements.

6.41. Document requirements.

6.42. Acquire bids.

6.43. Select equipment suppliers.

6.44. Define layout for equipment and utilities.

6.45. Design and construct special tooling, fixtures.

6.46. Design product production routing layout.

6.47. Acquire any required permits.

6.48. Order tools, machines, equipment, and utilities installation.

6.49. Run acceptance and capability testing.

6.50. Install and test hard tooling.

6.51. Install standard type facilities (desk, workbenches, storage racks, production line, computer systems, etc.).

6.52. Ensure all measurement equipment is calibrated, has a calibration procedure, and is on the calibration recall list.

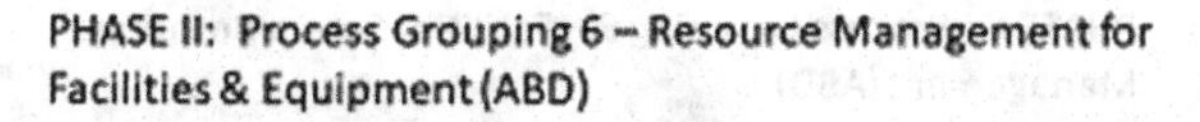

PHASE II: Process Grouping 6 – Resource Management for
Facilities & Equipment (ABD)

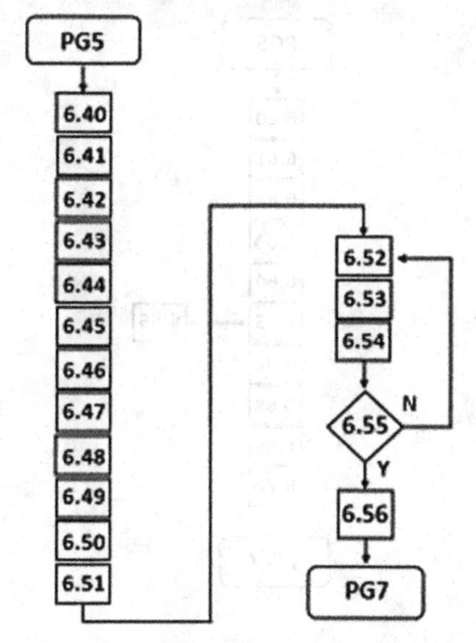

FIGURE 7.5
Process Grouping 6. Equipment Resource Management Activity Block Diagram

6.53. Make sure that all equipment, tools, and fixtures have been maintained in keeping with the maintenance instructions.

6.54. Equipment and utilities ready for pilot run.

6.55. Was pilot run successful?

6.56. Release for production when the production-related paperwork is complete and signed off and knowledge warehouse has been updated.

Forward to PG7.

Process Grouping 6: Facilities Setup Resource Management Activity Block Diagram

The following is the rest of a list of typical activities that take place during Process Grouping 6. Facilities Setup Resource Activity Block Diagram (see Figure 7.6).

6.60. Industrial engineering reviews space requirements defined in Business Case Analysis and updates as required.

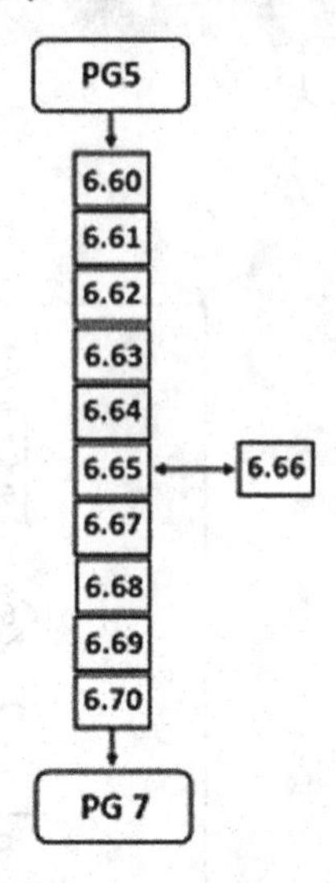

FIGURE 7.6
Process Grouping 6. Facilities Setup Resource Management Activity Block Diagram

6.61. Determine if the product lends itself to rearranging the production area for each different style entity.

6.62. Define environmental conditions requirements (temperature, humidity, air quality, electrical and audio noise containment, security, etc.).

6.63. Review property layouts and limitations.

6.64. Review to determine availability of space, utilities, layout, and equipment.

6.65. For all special machines and equipment, get the machine diagram to determine that the power connections are properly mounted on the floor and that consideration be given for the conveyor equipment.

6.66. If space is not available, additional outsourcing should be considered.

6.67. If there is still a problem, determine if rental or construction is preferable.

6.68. Review alternatives with the executive sponsor, agree on an action plan, and update the knowledge warehouse.

6.69. If the action plan cost exceeds the approved budget process, a project variance request needs to be approved.

6.70. If budget is approved, go to PG7.

Output goes directly to PG7.

(Note: There are many other options to solving the facilities limitation problem than the ones previously mentioned. Some of the other approaches could be: working a second shift to make better use of space, lower output expectations, use of home office, reduce output quantity schedule, move inventory storage off-site, etc.)

Process Grouping 6: Financial Resources Management Activity Block Diagram

Availability of human resources is equally as important to the success of a potentially innovative project/program as the availability of the required financing. In startup and small organizations, lack of sufficient financial resources turns out to be a major roadblock to the project success. Even in large successful organizations, its financial resources must be addressed and efficiently managed. Basically, when we're talking about implementing a potentially innovative project/program, the organization is determining how they will invest their discretionary financial resources.

- Definition of Discretionary Financial Resources: Discretionary financial resources are all of the financial resources that are available when the cost to maintain the present activities is subtracted from the total available financial resources. Discretionary financial resources are typically used to develop new products; pay bonuses; expand facilities, research and development, improvements made to the present organization; and acquire new businesses. Typical items included in maintaining the present activities would be finances required to support presently approved projects/products, pay interest on outstanding debts, maintain current facilities, materials costs, supplier costs, current production/manufacturing costs, maintenance and repair costs, employee salaries, pay dividends, etc.

The following list is of typical activities that take place during the Financial Resource Management activities (see Figure 7.7).

6.80. Define present financial resources committed by the organization. This includes present total operating costs plus financial commitments that will be funded during the innovative entities lifecycle. This data is

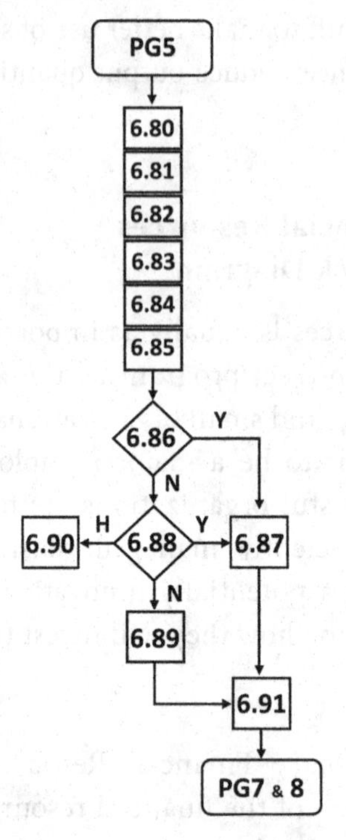

FIGURE 7.7
Process Grouping 6. Financial Resource Management Activity Block Diagram

known as Projected Financial Obligations or present financial obligations to contain current commitments.

6.81. Define current projected financial revenue/income and the organization's present reserve funds.

6.82. Define present financial obligations to contain current commitments.

6.83. Define the amount of financial resources that would be classified as discretionary financial resources. Subtract 6.80 and 6.82 from 6.81.

6.84. Using the priority order defined in Chapter 6 for the project or programs, process the approved projects/programs through the following financial resource assignment steps in priority number order.

6.85. Based on its priority, select the project/program to define its financial status.

6.86. Is there sufficient discretionary fund available to support the project?

6.87. Yes – Modify and approve budget if appropriate.

6.88. Look for sources of other funding. There are a number of ways to additional financial reserves. One of the most popular ways is to borrow money from a reputable bank or lending organization. Another is transferring stock into reserve funds by selling it. Many new organizations will look immediately for angel funding. Another very popular way is to implement an initiative to increase sales of present product or to slow down some of the other programs/projects that are presently approved.

6.89. No – No other financial resource options available.

6.90. Put on hold to be considered for future date.

6.91. Update the knowledge warehouse and terminate the project plan.

Go to PG 7.0.

PROCESS GROUPING 7: DOCUMENTATION

Everybody feels there's too much documentation that they have to do. We are all provided with far more information than we can possibly ever use and much more than we often would want. Still we are dependent on it to make our decisions. Our paperless office has been replaced with computers that end up providing us with much more data and information than we could possibly read, let alone comprehend. Unfortunately, the project cycle required to transform an innovative opportunity into a value-added reality requires a great deal of communication throughout the entire organization. The bigger the organization, the more documentation that is required. The more and more we learn, the more people we want to share our knowledge with. The more complex the things we are doing, the more need to have documentation in order to adequately manage the knowledge and information.

Document control and document management are key problems that the world faces today. Fortunately, today documentation does not necessarily mean mounds and mounds of paper. Most of our documentation is now stored way out there someplace in the cloud and is delivered to us at the touch of a key. Certainly, in industry, our focus on optimizing our business processes requires everyone to perform a similar activity in the same standard way. Unfortunately, transforming a fuzzy opportunity into a value-added reality requires a great deal of standardization. In many cases, documenting the systems required to make the transformation from a caterpillar (improvement opportunity) into a butterfly (value-added innovative item) is one of the most difficult and resource-consuming activities in the Innovative New Entity Cycle. Typical documents that are developed can be found in your organization's business procedures manual, engineering specifications, productions routings and instructions, security manual, safety manual, and sales and marketing instructions.

Six Document Management Systems

For this analysis, we have established only six document management systems. We are going to discuss the following:

- Part 1 Product Specifications Documentation
- Part 2 Project Management Plan and Facilities Planning Documents
- Part 3 Suppliers and Contractors Documentation
- Part 4 Production Setup Documentation
- Part 5 Producing Output Controls Documentation
- Part 6 Marketing and Sales Documentation

Outputs from Process Grouping 6. Resource Management are a project plan and work breakdown structure. All provide relevant information related to the specific project that the team is working on. It's important to understand that there is a great deal of cross play between Process Groupings 6 and 7. For example, there is very little manufacturing engineering can do until product engineering documents the product. Fortunately, in most organizations, there is close harmony between product engineering and manufacturing engineering, allowing manufacturing engineering to make decisions based upon discussions and observations long before the requirements are documented by product engineering. Excellent communication and cooperation between organizations prior to formal documentation being available is the secret to reduce new product development time.

The major functions involved in creating product documentation include product engineering, project management, manufacturing/test engineering, production control, industrial engineering, quality assurance, sales, and marketing.

Process Grouping 7: Part 1 – Product Specifications Documentation Activity Block Diagram

The following Product Specification Documentation Activity Block Diagram provides a view of some of the more critical documentation required to support a project for a new innovative product (see Figure 7.8).

Here is a description of the individual activities that make up this activity.

7.1.0. Establish a document control center. All documents, except individual personal records, flowing through the center distribution are controlled by the center confidential communications.

7.1.1. Review relevant information in the knowledge base.

7.1.2. Review model design configuration and performance.

7.1.3. Share key points related to the present program/project in personal engineering notebooks.

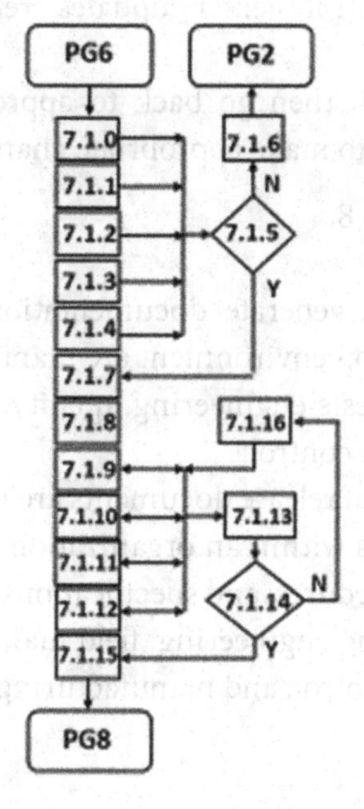

FIGURE 7.8
Process Grouping 7. Part 1 Product Specifications Documentation Activity Block Diagram

7.1.4. Using tests model data, compare relevant actual performance parameters to customer desired performance parameters.

7.1.5. Did it meet requirements?

7.1.6. No – Send back to Redesign Process Grouping 2.

7.1.7. Finalize and release product performance specification's required design review.

7.1.8. Release final output prints and specifications required design review.

7.1.9. Release assembly prints and specifications required design review.

7.1.10. Release component prints and specifications required design review.

7.1.11. Release installation documentation/validation required design review.

7.1.12. Release maintenance and repair documentation required design review.

7.1.13. Document approval is required by responsible functions before release.

7.1.14. Was release approved?

7.1.15. Document control center updates records and distributes documents.

7.1.16. If not approved, then go back to appropriate release activity (Activities 7.1.7 to 7.1.13) to make appropriate changes to 7.1.13.

Outputs – Process Group 8.

Typical functions that generate documentation required to support manufacturing/production environment are manufacturing engineering, industrial engineering, tests engineering, quality assurance, purchasing, shipping, and production control.

Note: Most of the formal release documents are required to be approved by a number of functions within an organization. For example, typically products' drawings, procedures, and specifications need to be approved by marketing, manufacturing engineering, field maintenance group, quality assurance, production control, and manufacturing.

Process Grouping 7: Part 2 – Project Management Plan and Facilities Planning Documentation Activity Block Diagram

The following is a description of the individual activities that make up the Part 2 Project Management Plan and Facilities Planning Documentation Activities (see Figure 7.9).

7.2.1. Develop a document identification numbering system and establish documentation formats primarily focusing on identification information. For companies with over 10 employees, a computer system is usually used.

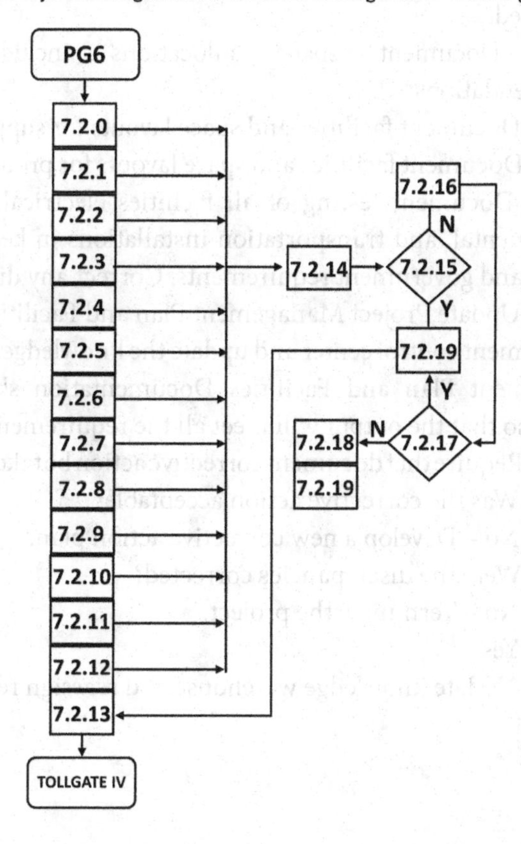

FIGURE 7.9
Process Grouping 7 Part 2 Project Management and Facilities Planning Documentation Activity Block Diagram

7.2.2. Establish rules for when hard or soft copies of documents would be kept on file.

7.2.3. Establish filing control and obsolescence control system.

7.2.4. Release formal procedures and train relevant staff on how to use the document control system.

7.2.5. Project team should check to ensure the product requirement system specifications (e.g., functionalities, speed, safety, and standards) are in the system and at the correct change level.

7.2.6. Input the project management plan and work breakdown structure into the control center database.

7.2.7. Submit scope of work for each project that includes budgetary numbers and the goal of each project.

7.2.8. Define the parts of the innovative entity production that will be outsourced.

7.2.9. Document space allocations including outsourcing recommendations.

7.2.10. Document facilities and space layouts for support staff.

7.2.11. Document facilities and space layouts for production and storage.

7.2.12. Document testing of all facilities electrical, water, chemicals, environmental, and transportation installations in keeping with organizational and government requirements. Correct any discrepancies.

7.2.13. Update Project Management Plan and Facilities Planning part of the document control center and update the knowledge warehouse. Project Management Plan and Facilities Documentation should be complete enough so that the output will meet all the requirements of Tollgate IV.

7.2.14. Require that document corrective action be taken on discrepancies.

7.2.15. Was the corrective action acceptable?

7.2.16. No – Develop a new corrective action plan.

7.2.17. Were the discrepancies corrected?

7.2.18. No – Terminate the project.

7.2.19. Yes.

7.2.20. Update knowledge warehouse and reassign resources.

Process Grouping 7: Part 3 – Suppliers and Contractors Documentation Activity Block Diagram

The following activities are typical of those that take place in order to provide the Process Grouping 7. Part 3 Suppliers and Contractors Documentation to support an approved project (see Figure 7.10).

Assign a team to develop the organization's supplier and subcontracting system and its documentation. Normally setting up the system is not considered a part of the project or program activities. It usually is a completely separate project unto itself, and the project team has a responsibility to follow the development process.

7.3.1. Document supplier/contract certification procedures.
7.3.2. Document supplier/contract reporting procedures.
7.3.3. Certify suppliers and subcontractors.

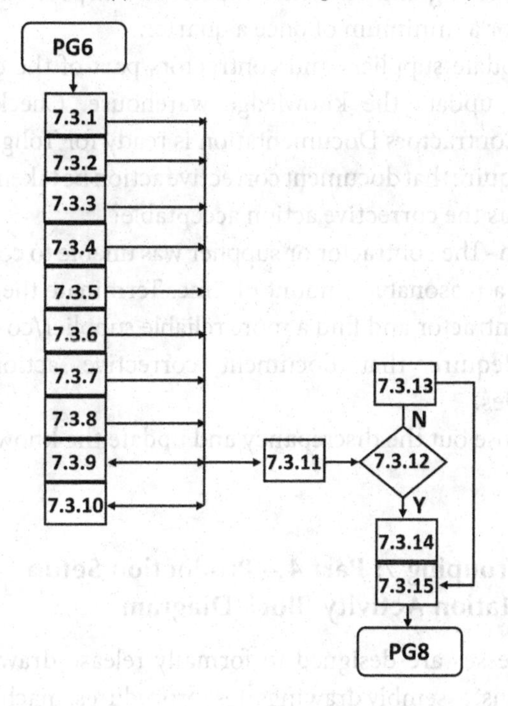

FIGURE 7.10
Process Grouping 7. Part 3 Suppliers and Contractors Documentation Activity Block Diagrams

7.3.4. Prepare inspection procedures for raw materials, components, and contracted items.

7.3.5. Plan and document the most profitable places to locate manufacturing or processing areas.

7.3.6. First part evaluation – Total inspection of supplier first items submitted. I personally also do the same thing on first lot submitted with a bigger sample. Often the pilot model that the supplier sends in is prepared by craftsmen that are often much more skilled and not working with the objective of getting a specific number of parts out per hour. For example, the chief chef may prepare the Lobster Newburgh the first time, but an apprentice cook will prepare all future servings.

7.3.7. Certify suppliers and order initial stocking of components, materials, and assemblies.

7.3.8. Document contracts with contractors, suppliers, and transporters.

7.3.9. Keep a continuous updated record of delivery date versus scheduled date, costs versus estimated, nonconformities as received, and defect rates during processing and customer experience. Report back to the supplier or contractor a minimum of once a quarter.

7.3.10. Update suppliers and contractors part of the document control center and update the knowledge warehouse. Check to be sure the Suppliers/Contractors Documentation is ready for Tollgate IV.

7.3.11. Require that document corrective action be taken on discrepancies.

7.3.12. Was the corrective action acceptable?

7.3.13. No–The contractor or supplier was unable to correct the reported incident in a reasonable amount of time. Terminate the contract with the supplier/contractor and find a more reliable supplier/contractor.

7.3.14. Require that document corrective action be taken on discrepancies.

7.3.15. Close out the discrepancy and update the knowledge database.

Process Grouping 7: Part 4 – Production Setup Documentation Activity Block Diagram

These processes are designed to formally release drawings, engineering specifications, assembly drawings, test procedures, machine set, inspection procedures, and complement specification formal release of production routing. The following is a description of the individual activities that

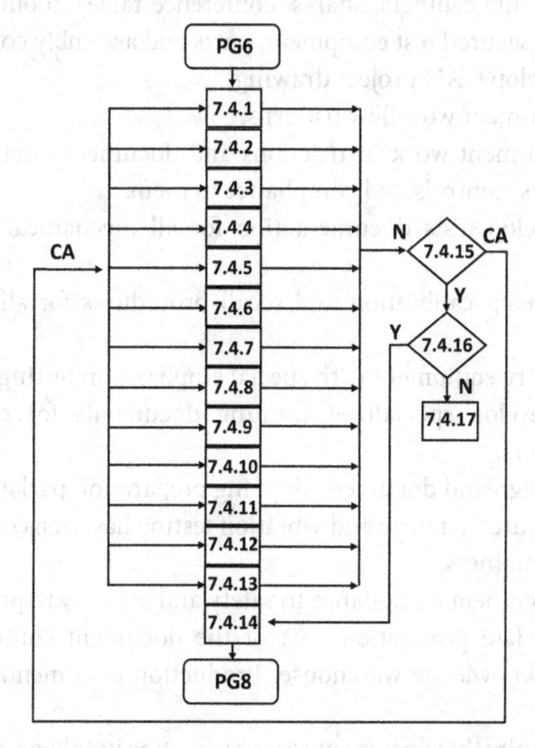

FIGURE 7.11
Process Grouping 7. Part 4 Production Setup Documentation Activity Block Diagram

make up the Production Setup Documentation Activity Block Diagram (see Figure 7.11).

7.4.1. Establish a team to document engineering change-level control process controls, document level controls, documentation format, identification structure, and the associated procedures.

7.4.2. Products/service production documentation needs to be available in the document control center (drawings, engineering specification, assembly drawings, and specifications). The project team will prepare a project plan that will define which of these documents need to be customized to the specific project/program, who will prepare them, when they will be prepared, how they will be approved, and how they get distributed and controlled.

7.4.3. Define and order equipment and facilities (computers, desk, workstation, file cabinets, shares, conference tables, tooling, production line, robots, secured test equipment, parts and assembly containers, etc.).

7.4.4. Develop CAD project drawings.

7.4.5. Document workflow patterns.

7.4.6. Document work instructions and document safety and security requirements, controls and compliance to them.

7.4.7. Develop tests documentation for all mechanical and electrical systems.

7.4.8. Develop calibration and recall procedures for all measurement equipment.

7.4.9. Certify equipment with special emphasis on testing equipment.

7.4.10. Develop specialized training documents for employees and subcontractors.

7.4.11. Design and document shipping preparation packaging.

7.4.12. Be sure that drop and vibration testing has been completed on the shipping containers.

7.4.13. Document compliance to safety and security requirements.

7.4.14. Update production part of the document control center and update the knowledge warehouse. Production documentation ready for Tollgate IV.

7.4.15. Require that document corrective action be taken on discrepancies.

7.4.16. Was the corrective action acceptable?

7.4.17. Terminate project.

Process Grouping 7: Part 5 – Producing Output Controls Documentation Activity Block Diagram

Document production reporting data systems and analyze production processes, schedules, methods, and other data. Then provide management with reports containing the data and statistics to enable management to better understand future requirements needed for the manufacturing process (see Figure 7.12).

7.5.1. Production control prepares a production schedule for the entity.

7.5.2. Procurement releases purchase orders for parts and assemblies.

7.5.3. Work orders are prepared for the individual production departments involved in the entity.

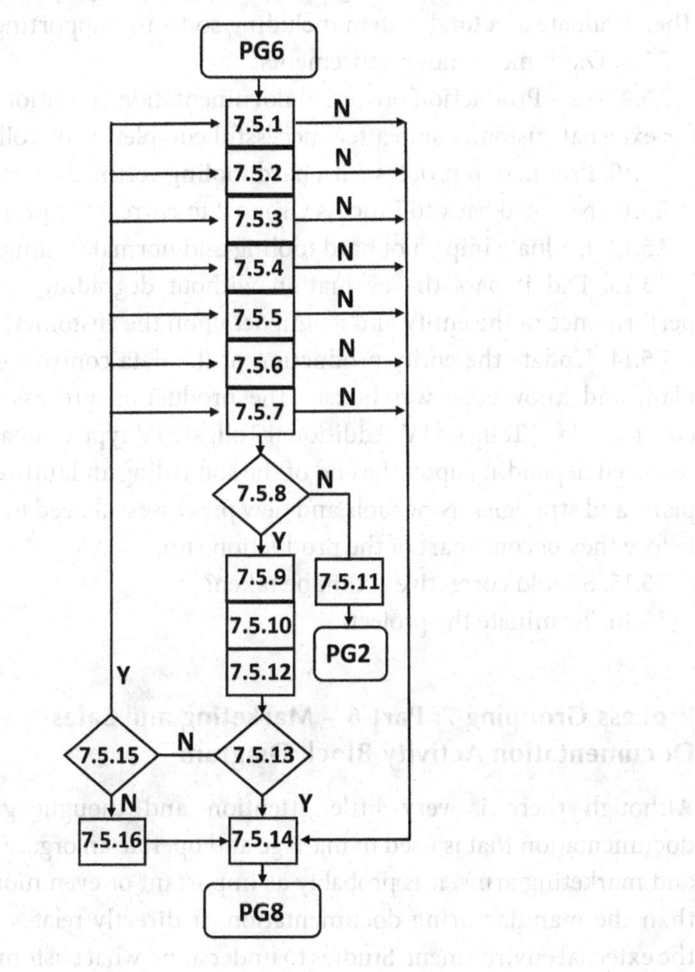

FIGURE 7.12
Process Grouping 7. Part 5 Producing Output Controls Documentation Activity Block Diagrams

7.5.4. Employees are trained to perform specific parts of the process. Often certification is required.

7.5.5. Develop certification compliance processes and document (e.g., ISO 9000).

7.5.6. Perform initial and ongoing design reviews and make changes as required.

7.5.7. Conduct individual segments of the entity system pilot run with temporary and/or hard tooling with permanent or temporary personnel. Then evaluate as a total system including software supporting activities.

7.5.8. Did it meet pilot requirements?

7.5.9. Yes – Production processes documentation conditionally certified for external customer ship after successful completion of Tollgate IV.

7.5.10. Production process with hard tooling certified.

7.5.11. No – Go back to Process Group 2 to correct the problem.

7.5.12. Evaluate impact of hard tooling and normal staffing.

7.5.13. Did it pass the evaluation without degrading the quality or performance of the entity and its impact upon the customer?

7.5.14. Update the entity production of the data control center, project plan, and knowledge warehouse. The production process can now be considered for Tollgate IV. Additional Tollgate IV type evaluations may be required depending upon the end of the soft ruling and future automation plans and strategies. New tools and new processes all need to be evaluated before they become part of the production run.

7.5.15. Should corrective action be taken?

7.5.16. Terminate the project.

Process Grouping 7: Part 6 – Marketing and Sales Documentation Activity Block Diagram

Although there is very little attention and thought given to the documentation that is used to manage and operate an organization's sales and marketing arms, it is probably as important or even more important than the manufacturing documentation. It directly relates and impacts the external environment. Studies to understand what customers need and what their future needs will be are extremely important to successfully managing a business. Sales campaigns and marketing strategy can do more to making a product successful or a complete failure than cosmetic defects in the end product. The two areas where the most creative people exist in your organization is usually product engineering and sales and marketing. Organizations recognize the important contribution the sales and marketing team makes by paying in many cases significantly more than they pay in engineering. Someday, some bright young quality engineer will establish a system where the quality organization can contribute to the quality of the sales process, delivery, and after-sales service.

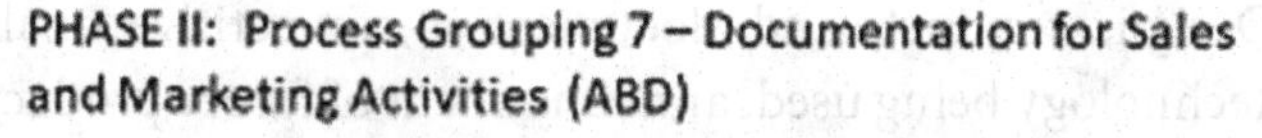

PHASE II: Process Grouping 7 – Documentation for Sales and Marketing Activities (ABD)

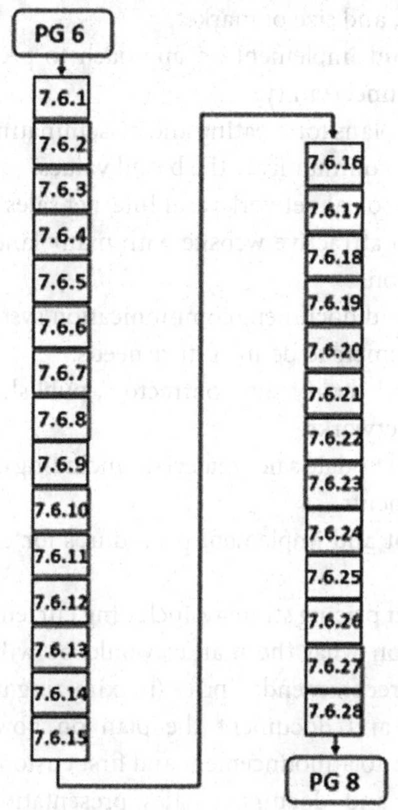

FIGURE 7.13
Process Grouping 7. Part 6 Marketing and Sales Documentation Activity Block Diagram

The following are the typical activities that take place during Process Grouping 7. Marketing and Sales Documentation (Figure 7.13).

7.6.1. A project team should be assigned to prepare the processes and procedures related to how a project is handled in marketing and sales and the documentation required to support a new innovative entity from the marketing and sales standpoint. This is normally not done by the project team assigned to an individual entity. This is normally done by a project team that has these procedures and documentation assignment as the team objective.

7.6.2. Create customer research and survey to provide information relative to enhancing or redirecting sales/marketing activities in the product type.

7.6.3. Design, conduct, and document a competitive analysis that includes technology being used, announcement of future product, pricing of current product, and size of market.

7.6.4. Develop and implement an approach to predict the size of the market with a low uncertainty.

7.6.5. Document plans for creating and disseminating images, messages, and ideas that best communicate the brand values.

7.6.6. Document social networks and Internet sales strategies.

7.6.7. Develop an attractive website with many landing pages that lead to more conversation.

7.6.8. Establish and document communication systems with customers and potential customers to define future needs.

7.6.9. Identify and certify sub-contractors, publishers, designers, radio stations, and TV networks.

7.6.10. Create and update sales materials, including catalogs, advertising, and TV advertisements.

7.6.11. Document and implement procedures for controlling customer contact.

7.6.12. Document pricing strategy, including current pricing of product/services based upon what the market would be willing to pay and the market size at the recommended price (maximizing income).

7.6.13. Develop and document the plan on how to handle special promotions related to announcement and first customer ship.

7.6.14. Develop and document sales presentations and visual aids strategy.

7.6.15. Prepare and publish news releases related to the new product.

7.6.16. Prepare and begin advertising activities.

7.6.17. Document and implement a procedure to handle client referrals.

7.6.18. Develop and document customer retention techniques (e.g., customer relation management software packages).

7.6.19. Develop and maintain maps.

7.5.20. Develop and maintain a list of qualified leads.

7.6.21. Develop sales proposal.

7.6.22. Prepare and present sales contracts.

7.6.23. Model all major processes and instructions.

7.6.24. Distribute all documentation as appropriate based upon security and organizational direction.

7.6.25. Prepare your standard sales pitch for customer service center.

7.6.26. Prepare training materials for field service personnel and customer service center.

7.6.27. Prepare a position paper to be considered during Tollgate IV.

7.6.28. Update knowledge management at regular intervals.

Output goes to PG 8.

Summary Process Grouping 7: Documentation

During this activity, the engineering documentation, maintenance manuals, production routings, and job instructions are prepared and operators are trained on how to use them. Packaging and shipping containers are evaluated to ensure that they provide adequate protection for the product. The information collection system is defined and put in place. The project management data system generates frequent status reports to keep the management team aware of the status and point out activities where they need to be involved in.

PROCESS GROUPING 8: PRODUCTION

The organization has spent a great deal of money to develop the entity to this point in the innovation cycle. Unfortunately, it is all negative cash flow, and we still have a great deal more to invest before we can start to earn back the money we invested and just a little bit more to pay for the use of the investor's money. The time is right to emphasize the importance of bringing up the process capabilities to the point that the delivered output meets or exceeds all of the product requirements and the salesperson's promises. Your first new innovative items shipped to an external customer should be at least as reliable as the item it's replacing, even though its performance capability is far superior. It is production that transforms wishes into dreams, hopes into goals, and an output that a customer will line up to buy whether they really need it or not. It's that magical point in the innovative cycle that transformed an empty echoing room into a bustling noisy, exciting, and productive environment where people, machines, and materials transform into items that support the wonderful lifestyle we enjoy here in the United States.

I cherish our production activity for without it there is nothing. I liken it to printing a newspaper. The presses start to roll slowly at first pulling the reluctant paper through as the print as nestles in its own little pocket. In seconds, the row changes to a hum, and in just a few minutes, the hum changes into a rhythmic song of its own as a paper seemingly flies through the rollers cutting and folding up into the piece of paper that paints a picture of our complex world for all of us to see and understand.

Process Grouping 8: Production Activity Block Diagram

Process Grouping 8. Production Activity Block Diagram provides a view of some of the more critical activities required to support a project for a new innovative product. The following is more detailed descriptions of the individual activities that make up the Production Activities (see Figure 7.14).

8.1. Review project management plan and update if appropriate.
8.2. Review knowledge database.
8.3. Review output delivery schedule.
8.4. Review production cycle time estimates.
8.5. Characterize the production area. Be sure you know exact dimensions and shape of the areas that have been set aside to protect the entity.
8.6. Are there any changes that need to be made in the plan for layout?
8.7. Update work area layout and product flow.
8.8. Evaluate airflow, waste support disposal system, and water to be sure that it will be capable of handling the entity that is being produced in the area (e.g., presses air conditioning, contamination, quantities being processed, used chemicals).
8.9. Install equipment as it is delivered and test it to be sure that it is functioning properly. Often, this is best done by the company you're buying the equipment from.
8.10. Install hard tooling (e.g., dies, fixtures, punches, ovens, washers, safety equipment) and run an acceptance test on each one individually.
8.11. Connect everything together with the conveyor belt and storage cabinet that operates the system.

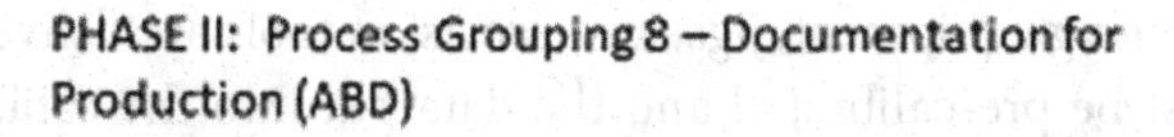

PHASE II: Process Grouping 8 – Documentation for
Production (ABD)

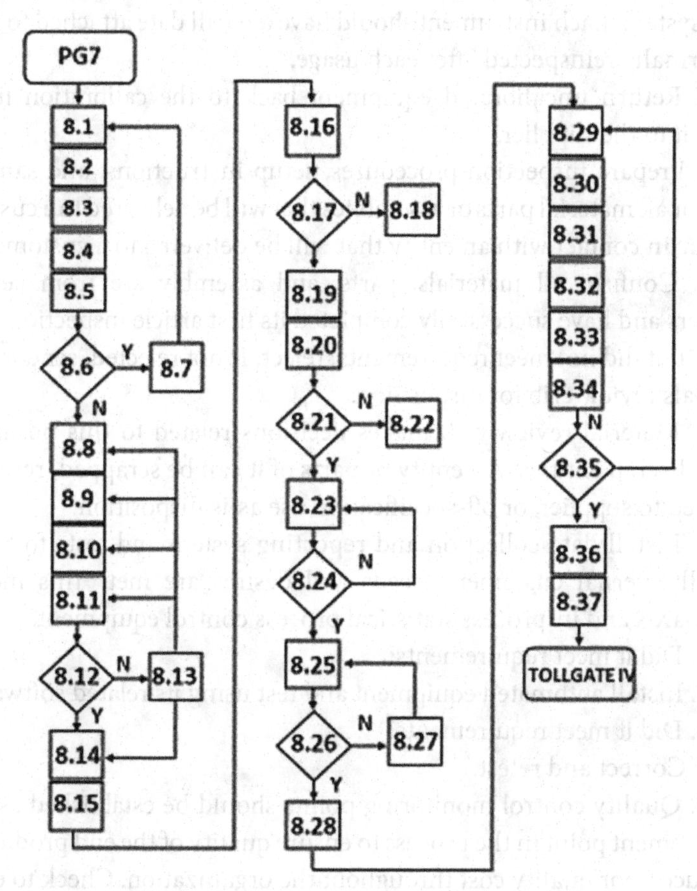

FIGURE 7.14
Process Grouping 8. Production Activity Block Diagram

8.12. Did the facilities function properly?

8.13. Modify as necessary.

8.14. Operates as required.

8.15. Experiment with each major piece of equipment to determine proper setup and operating instructions with particular attention to the tool cutting angles and a set of procedures.

8.16. Prepare setup instructions for equipment including maintenance procedures. Make sure that all equipment, tools, and fixtures have been maintained in keeping with the maintenance instructions.

8.17. As measuring equipment, jigs, and precision tools are delivered to the area, it must be pre-calibrated and the data put into the calibration recall system. Each instrument should have a recall date attached to it. Jigs are normally reinspected after each usage.

8.18. Return uncalibrated equipment back to the calibration line or return it to the supplier.

8.19. Prepare inspection procedures, setup instructions, and sampling plans for all material parts or assemblies that will be delivered to a customer or come in contact with an entity that will be delivered to a customer.

8.20. Confirm all materials, parts, and assembly are from certified suppliers and have successfully completed its first article inspection.

8.21. If it did not meet requirements, reject. If not rejected, send it to the materials review crib for disposition.

8.22. Material review crib makes decisions related to this position of rejected/scrap entities. An entity or parts of it can be scrapped, reworked, returned to supplier, or off-specification use as-is disposition.

8.23. Install data collection and reporting system and tests to be sure that all internal customers' needs and desires are met. This includes dashboards and in-process statistical process control equipment.

8.24. Did it meet requirements?

8.25. Install automated equipment and test using its related software.

8.26. Did it meet requirements?

8.27. Correct and retest.

8.28. Quality control monitoring points should be established as a key measurement point in the process to ensure quality of the end product and to reduce poor quality cost throughout the organization. Check to ensure that quality and cost are being measured and reported.

8.29. Training documentation and classes should be held to prepare workers who come in contact with the deliverable entity to be sure they have been trained in their roles and responsibilities.

8.30. Manufacturing engineering should make pilot runs until they are confident that everything is operating correctly and equipment and utilities are ready for pilot run for producing customer acceptable product that meet all the performance requirements.

8.31. Finalize packaging design as a product that would be presented to the consumer and obtain product packaging materials for use in the production line. This is a very important design feature as it can have a major impact upon sales. We have seen occasions where the packaging has a bigger impact upon selling the product than the entity needed inside the package.

8.32. Select shipping materials and test to ensure no damage will occur under normal handling and shipping conditions.

8.33. Make certification runs using qualified employees paying particular attention to the distribution at each critical measurement point.

8.34. Did it pass a certification run?

8.35. Release for production when the production-related paperwork is complete and signed off.

8.36. Activate the process with individuals who will normally produce the entity making product that may or may not be customer shippable based upon the results of Tollgate IV.

8.37. Update the knowledge warehouse (database).

Output goes to Tollgate IV.

TOLLGATE IV: CUSTOMER SHIP APPROVAL

At this point in the process, the entity design should be complete, utilities equipment and space should be set up and operational, and employees should be trained and functioning in the process in accordance with their long-term assignment. The entity being produced should meet all of the engineering, sales, maintenance, legal, and consumer requirements. The last activity in this sequence is an evaluation to determine that all the conditions required to produce the entity in the projected quantities have been complied with and that the final entity represents the reputation that the organization wants to achieve. This very thorough evaluation needs to be conducted using the normal operating equipment and processes that will be used to produce the entities that go to the external customer/consumer. This evaluation is called Tollgate IV. Customer Ship Approval.

Important note – Customer-centric organizations are very careful to ensure that this Tollgate assessment confirms that the products the organization is shipping are representative of the organization's values, beliefs, and commitments.

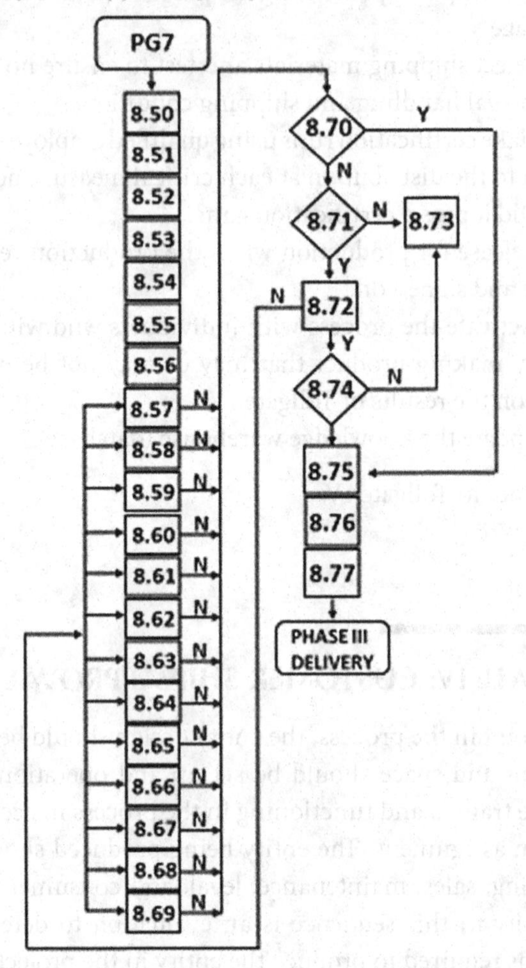

FIGURE 7.15
Tollgate IV. Customer Shipment Approval Activity Block Diagram

Tollgate IV: Customer Ship Approval Activity Block Diagram

The following is more detailed descriptions of the individual activities that make up the Tollgate IV. Customer Ship Approval Activities (see Figure 7.15).

You'll note that we are using the same first digit (8) that we use for the Production Activity Block Diagram because of their close interrelationship and because Tollgate IV activities are closely related to evaluating how effective Product Grouping 8 – Production was performed.

8.50. Establish a Tollgate IV team to conduct the evaluation and make recommendations to the executive team. Often this team is led by the quality assurance function.

8.51. Make sure the team is familiar with the entity specifications and the information related to the entity that is contained in the knowledge management system.

8.52. Develop a Tollgate IV assessment plan that is approved by related functions.

8.53. Define data collection points, means of collecting data, who will be collecting the data, how will the data be recorded, and how will it be analyzed. It is very important that cycle time, processing time at each operation, and yields per operation information be collected in order to determine process output capabilities. Be sure that the plan includes collecting information related to potential customer comments during the pre-customer ship sales and marketing campaign.

8.54. Define the sample size of the entities that will need to be processed in order to get sufficient confidence that the entity is ready to be shipped to a customer/consumer and that the organization has the capabilities of producing the projected quantities. Many of the bad decisions that are made by the executive team are the result of them being presented with data that often has as much as a 50% potential error in the projection (e.g., the project will take 12 months to complete, potential variance from estimate +16 months or minus one month).

8.55. Determine if a control sample needs to be run in parallel with the entities used to evaluate Tollgate IV. We have seen cases where the additional attention given to a process area when it is being redesigned improves the original process more than the new innovative proposed process. With a high percentage of innovative changes being patented are minor changes to an already-established process, it is good practice for running a control sample along with the new proposed innovative entity. We normally do not recommend making a change unless you can identify value-added content that results from the change activities.

8.56. Usually different kinds of data are required to do the Tollgate analysis that is normally collected to ensure the process is functioning properly. As a result, the individuals collecting the data need to be trained in how to use the equipment, the forms that are used, and how to input the information into the computer system.

8.57. With data collected evaluate if the data is acceptable and the entity and its processes meet requirements.

8.58. Is the process in compliance with required governmental standards (e.g., safety, fire, materials handling, legal, financial, patents inferences, etc.)?

8.59. How do you compare to industry standard (e.g., quality, verbal, safety, product, customer, standards, and best practices)?

8.60. Is everyone producing the output trained for the job they're doing and certified when defined as a requirement?

8.61. Are all the materials, parts, and entities that are to be used in the innovative entity coming from suppliers and contractors that are certified as acceptable suppliers/contractors for the specific item being delivered to external customers/consumers?

8.62. Are the employees performing the work in keeping with the work instructions and documentation, or are they doing it the way they think is best?

8.63. Is a computer reporting system in place that keeps appropriate levels of management informed about the status of their products in-house and in the field?

8.64. Is after-sale services activities operational with trained personnel ready to handle customer questions, complaints, and suggestions?

8.65. Is design floor control in place and working?

8.66. Are the items packaged and shipped in packaging appropriate for ease of handling and to prevent damaging during shipping and unpacking?

8.67. Is the space layout and equipment plan adequate to meet peak production demands?

8.68. Are the procedures related to training the sales and marketing team? Do they feel maintenance team is adequate, and have the individuals been trained in keeping with the plan? Note: After-sales maintenance and service starts when the first customer delivered item is sold. A field-trained sales and after-sales service staff should be set up before the first entity is sold to its external customer.

8.69. Are the sales materials in agreement with the engineering specifications and announced performance abilities?

8.70. Were all the requirements of Tollgate IV met?

8.71. Should corrective action cycle start?

8.72. Yes – The situation is returned to Process Grouping 8. Production to evaluate the discrepancies and take corrective action. The Tollgate team will evaluate the effectiveness of the implemented corrective action. This usually requires additional test data.

8.73. No – Management will decide if the project will be dropped, put on hold, or modified. Often this occurs after corrective action has been implemented but is proven not to be effective.

8.74. Was corrective action successful?

8.75. Yes – Prepare Tollgate IV report and present the team's findings to the appropriate management at a meeting. As a result of this meeting, the entity may be classified as being customer shippable. In some cases, the management team may require additional information or preventive action before a final decision can be made. In these cases, Tollgate IV plan is adjusted to meet the management team's needs, and the final report is modified to reflect this additional information. At this point of time, the floodgate should be open and product should be being delivered to external customers/consumers.

8.76. The preliminary Tollgate IV report is usually presented to the appropriate managers at a meeting.

8.77. Update the document control center and the knowledge warehouse.

The output entities are now ready to be shipped to external customers/consumers and the project advances to Phase III. Delivery production documentation ready for Tollgate IV.

Summary Process Grouping 8: Production

As soon as the product is approved for shipment to the customer/consumer, the manufacturing floodgate is opened. The documentation and estimates are put under stress to meet the initial output demands that occur at the start. The information collection system is initialized and status reports are generated.

At this point in the project, we are deeply committed to our customers and our investors to continue and make the project a success. Typically, with manufactured products, services, and software, we have already committed orders from external customers. From an investor standpoint, a great deal of the organization's resources has already been expanded with no return on investment. Money that resulted in bigger dividends was redirected to funding the development of this opportunity. To stop shipment at this point of time would require the output to not perform at the level committed to the external customer or a safety problem. As a result, this Tollgate is primarily directed at ensuring that all the "i's" are dotted and all the "t's" are crossed.

It's at this Tollgate we should compliment ourselves on doing a splendid job that we are proud of, ensuring that no small item slips through the crack. It goes beyond reviewing the output to ensure that it has a high probability of beating the advertised performance requirements and the customers' expectations. It also focuses on the support activities like warehousing, processing sales orders, supplier order tracking, after-sales service, spare parts quantities, etc. Its analysis is to ensure that everything that could be done has been done to provide the external customer/consumer with an experience that delights them and exceeds their expectations.

Typically Tollgate IV is conducted two to three weeks prior to the first-customer ship date so that any identified improvement opportunity can be implemented in the first-customer ship item.

> *For replacement products, the new items should be performing equal to or better than the obsolete last product that was produced. The new product's performance curve should be at the high point in the replacement part's performance curve.*
>
> **—H. James Harrington**

At this point in the cycle, the production facility should be hard tool and ready for producing the projected quantities of output in a normal 40 hours week. The performance of the product should be in keeping with or better than the customer/consumer expectations. It also should be better in performance quality, reliability, safety, and costs than the replacement item. Your customer should never be their final inspection operation.

8

Phase III: Delivery

INTRODUCTION

During this phase, the output from the process is transformed from items into dollars and cents. It also includes a performance analysis to compare actual results to project value added to stakeholders. In Phase III, there are four Process Groupings and one Tollgate.

- Process Grouping 9: Marketing, Sales, and Delivery
- Process Grouping 10: After-Sales Service
- Process Grouping 11: Performance Analysis
- Tollgate V: Project Evaluation
- Process Grouping 12: Transformation

PROCESS GROUPING 9: MARKETING, SALES, AND DELIVERY

Here we have entered a different world. Somehow the sales and marketing activities and culture are uniquely different from the culture in other parts of the organization. During this activity, promotional and advertising campaigns are developed and implemented. Sales strategy and quote approaches are prepared, and the motivational compensation packages are designed. Since key systems that are frequently overlooked until a problem occurs are accounts receivable and order fulfillment systems, it's important to understand that the sales cycle is not completed until the item has been received by the customer and the customer pays the organization for the product or service that they received.

DOI: 10.4324/b22993-8

During the documentation activities, the marketing and sales plans and strategies were developed. It is during this Process Grouping, the adequacy of the plans and strategies are evaluated, measured, and updated.

Your marketing plan should be directed at including the needs and expectations of the five Cs of marketing.

- Company
- Customer
- Competitors
- Collaborators
- Climate

The marketing plan should consist of the following 10 sections:

Table of Contents

1) Markets
   Research, study, and understand their current and future markets. Understand your customers' current needs, desires, and direction. Define products and services that will attract new customers while maintaining your present customer base.
2) Competition
   Use competitive intelligence to maintain and expand your market share. Understand the strengths and weaknesses of the competition's offerings through techniques like competitive disassembly analysis and functional testing. Use techniques like competitive shopping to understand competitive services advantages.
3) Distribution
   Work on developing and expanding distribution. Look for ways to collaborate with other businesses to decrease distribution costs and increase item demands. Focus on reducing cycle time between order entry and delivery.
4) Supply Chain
   Make sure that you are not at the mercy of the wholesaler so that you mark up the price without real justification. Certified backup suppliers provide a degree of risk reduction for critical and long-cycle time procured items. Continuously assess economics related to internal or external value-added content and costs.

5) Positioning
   Understand how you fit in the market, which provides the means of determining where your potential customers are, what is the right approach, and should you be using social media more extensively. Analysis of all sales is even more important than analysis of sales made.
6) Promotion
   Promotions help you define and zero in on specific demographics, thereby allowing you to reach more targeted customers. You need to understand the specific characteristics of the individuals using the various platforms.
7) Pricing
   Be sure the price you charge for an item is based upon the market, not based upon the costs to produce the item. There are some limitation guidelines that are valuable considerations. First, you don't want to exceed the upper side of the item price range in the market. You may want to vary prices based upon individual considerations (for example, if you're selling books, you may want to give a lower percentage decrease in price in November and December when buying is high for Christmas than you use the rest of the year).
8) Customer Service
   You lose more customers based on poor customer service than you do on inferior products. Customer service, in this case, includes the sales engagement, organization-provided installations, call centers, and order processing. Here is a good example. Suppose you call into the customer support call center for software packages and you are told the wait time is two hours because there is a heavy backlog. This might make you believe that there are a lot of problems with that product that is causing the backlog, so you decide not to purchase it.
   It's also important to seek feedback. Make sure you provide ways for customers to review your business. If you have a customer loyalty program, regularly contact customers who have not made purchases in a while to offer discounts or inquire about why they have not visited your business recently.
9) Financing
   I personally feel that the first obligation an organization has is to the individuals who finance the organization with a secondary

responsibility to the customer. For innovative products, the biggest concerns an organization will have is funding an item for a period of time when there is no return on investment. Continuously reviewing your capital structure is mandatory to maintain an operational organization. It's only when you become financially stable that are you viewed as a successful business in your community and able to build a solid reputation.

10) Consistent Strategies
    Don't treat your strategies and plans as cast in concrete. It's better to think of them as a ball of clay that you are continuous reshaping to stay aware of the business environment and the changing customer needs. Building customer loyalty and increasing sales is the way to a successful business that is rewarding to your stakeholders. This requires a continuing collection of knowledge that is reshaping your strategy and plans. In many organizations a two-year-old plan is more of a handicap than an advantage.

Sales strategy should be based upon the following five Ss of sales:

1) Sell – Gross sales. Sales are directly related to revenue.
2) Speak – Get close to customers. Communicate! Communicate! Communicate with your customers and potential customers. Know them as individuals as well as organizations.
3) Serve – Add the value. Focus on how your item adds value to their organization.
4) Save – Save costs. Be innovative to reduce your costs so you can reduce your item's costs to your customer. Find the most effective way to use your product at minimum cost and maximum value added.
5) Sizzle – Extend your brand positive recognition. Focus on maintaining present customers; it is less expensive than finding new ones.

Your marketing plan should include at a minimum the following seven sales items:

1) Sales messaging
2) Sales processes
3) Sales managing

4) Sales competency and behaviors
5) Sales adaption and execution plans
6) Sales enablement
7) Sales management development

Last, we have come to the part of the innovation product cycle where the organization has made a major investment developing and producing the new entity. By now the organization should have positioned itself to the point that the output should be viewed as value added to the potential customers. Basically, we have reached the point where the salesperson has the entity in hand that he can trade with the customer in return for something the customer has that the organization values more than the output they created. We like to think that innovation occurs when one party has something that is unique and different that they are willing to transfer to a second party for something that the first party precedes as being of more value to them than the entity they are transferring to the second party. Unfortunately, that does not agree with the definition of innovation documented in ISO standard 56000:2019. ISO Standard 56000 in clause 4.1.1 defined innovation as "a new or exchanged entity, realizing or redistributing value." (©ISO. This material is excerpted from ISO 56000:2020, with permission of the American National Standards Institute (ANSI) on behalf of the International Organization for Standardization. All rights reserved.)

Process Grouping 9: Marketing, Sales, and Delivery Activity Block Diagram

This Process Grouping consists of a group of processes designed to make this transfer of value between the two parties occur. Without this smooth transfer, you do not have an innovative entity. Typical activities that are included in marketing, sales, and delivery are defined in the following Activity Block Diagram (see Figure 8.1).

9.1. Create customer research and survey to provide information relative to enhancing or redirecting sales/marketing activities in the product type.

9.2. Design, conduct, and document a competitive analysis that includes technology being used, announcement of future products, pricing of current product, and size of market.

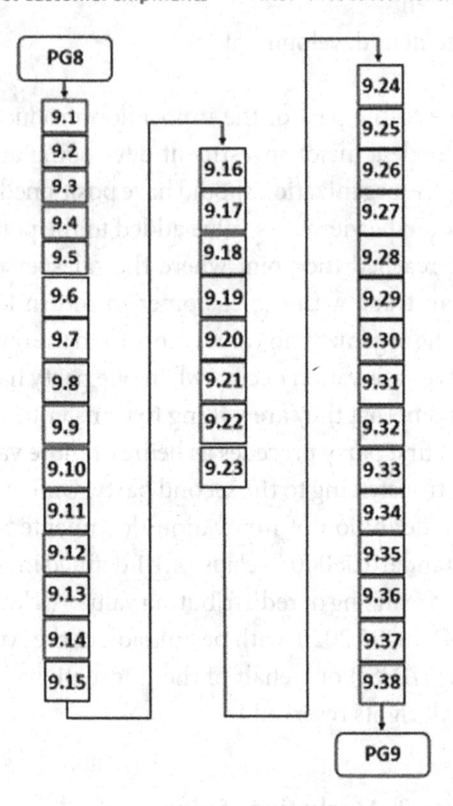

FIGURE 8.1
Process Grouping 9. Marketing, Sales, and Delivery Activity Block Diagram

9.3. Update project plan.

9.4. Document procedures for controlling customer contact.

9.5. Distribute all documentation as appropriate based upon security and organization direction.

9.6. Conduct initial entity-related training and performance specifications.

9.7. Develop and maintain empathy maps.

9.8. Document social networks and Internet sales strategies.

9.9. Document plans for creating and disseminating images, messages, and ideas that are brand values related.

9.10. Develop and implement an approach to predict the size of the market low uncertainty.

9.11. Develop an attractive website with many landing pages that lead to more conversation.

9.12. Identify and certify subcontractors, publishers, designers, radio stations, and TV networks.

9.13. Model all major processes and instructions.

9.14. Create and update sales material, including catalogs, advertising, and TV advertisements.

9.15. Install supporting software like customer relation management.

9.16. Develop unpacking and implementation procedures.

9.17. Interpret trade terms.

9.18. Develop and conduct entity-related and sales-related training for the marketing and sales personnel.

9.19. Prepare training materials and conduct classes for field service personnel and customer service center.

9.20. Document pricing strategy including current pricing of products/services based upon what the market would be willing to pay and what they will buy (maximizing income).

9.21. Select shipping approaches, suppliers, and packaging.

9.22. Generate leads across multi channels:

- Lead nurturing,
- Lead qualifications,
- Lead scoring,
- Lead prioritization,
- Lead development strategy.

9.23. Install a purchase order status reporting system after first customer ship.

9.24. Support for multi-LinkedIn member accounts.

9.25. Update the knowledge management system.

9.26. Generate competition awareness and knowledge.

9.27. Set and achieve volume requirements.

9.28. Set and achieve distribution objectives.

9.29. Make cold calls.

9.30. Document recommended sales presentations and visual aids.

9.31. Customize marketing and sales strategies/materials to meet unique sales opportunities.

9.32. Develop product displays and merchandising.

9.33. Prepare the standard sales pitch for customer service center.

9.34. Develop process to source and develop client referrals.

9.35. Develop and document customer-retention techniques.
9.36. Develop sales proposals.
9.37. Prepare and present sales contracts.
9.38. Close sale and schedule delivery.

The output from Process Grouping 9 goes directly to Process Grouping 10. After-Sales Services.

Summary of Process Grouping 9: Marketing, Sales, and Delivery

Most people relate innovation and creativity primarily to activities that take place in Product Engineering and R&D. I will agree that it is a hotbed of creativity. I felt exactly the same way until I took over as CEO of a midsize company and was forced to understand the entire innovative cycle. I personally believe there is more creativity and innovation required in the sales and marketing group than there is in the product engineering group. Promotional campaigns, dealing with individual clients, and the ability to readjust a sales campaign to make it more specific to a client, etc. require a great deal of creativity and ingenuity on the marketer and sales person. I guess that's the reason that in successful companies the sales and marketing personnel are paid much more than the development engineer.

Years ago we took our production employees off from the piecework-type pay, but we were still paying our salespeople based upon meeting their quotas. Certainly, working in sales and marketing is challenging, exciting, and rewarding as each sale is a win, and winning is the name of the game today. Today's marketers and salespeople live in a very fluid world where they have to be at faster than the markets they are servicing or else they will lose ground rapidly. Commission-based salespeople are highly motivated individuals because their success or failure primarily is based upon their creativity and ingenuity in dealing with their customers. I guess according to the definition of innovative as a new and different item that is produced meets the requirements to be called innovative. Personally, I feel that a new product that does not have a positive return on investment is not innovative in a for-profit company. If it doesn't have a positive return on investment, it may be a creative product, but I question if is it really innovative. I guess if we go with the ISO standard 56002, it would be considered innovative.

During this activity, promotional and advertising campaigns are developed and implemented. Sales strategy and quote approaches are prepared and the motivational compensation packages are designed.

PROCESS GROUPING 10: AFTER-SALES SERVICE ACTIVITIES

After-sales service includes individuals who man the control center, handle customer complaints, answer customer questions, and provide a line interface between the organization and its clientele. Another key part of after-sales service is a repair center. These two areas have to have the "patience of Job" since they are continuously faced with unhappy customers who just need someone to be mad at. Empowerment is the most useful weapon you can give these people.

After-sales service starts immediately after the customer and supplier have agreed to the terms related to the entity being acquired. The very first step in after-sales service is for the salesperson to explain the experience and the process that the customer will be subjected to in order for the entity to be put to use by the customer/consumer. This process needs to be well documented, and the people involved in the process need to be well trained and highly empowered to take whatever action is necessary to satisfy the customer. Studies have proven that customer satisfaction decreases by the square root of the number of individuals they have to discuss the problem with before a satisfactory win–win solution is developed.

Sales, marketing, and after-sales services are where the major part of an organization's creativity, innovation, and originality exist. The complexity of dealing with people overshadows the complexity of the most difficult engineering challenge. Each person an after-sales service employee comes in contact with has developed a set of emotions that is unique to his or her circumstances. To make it even more complex, these emotions change almost instantaneously, creating a whole new set of challenges and opportunities.

It's like trying to catch a cloud on a windy day. It is for these very reasons that the sales/service personnel must be much better trained. To relate a personal example, I have just completed writing a book made up of 19 short stories. This is the first fictional book I have written, and the publishers I have been using do not handle fictional-type books. So after a number of discussions, I decided to hire a firm that helps individuals self-publish a manuscript. I finally selected one that was highly recommended by a number of different organizational assessment firms on the Internet.

Over a period of eight days, we made four different phone calls to the publisher, and each time we went through the same process:

Step 1 – press 1, if you've got a problem with accounting.
Step 2 – press 2, if you want to find out the status of your book.
Step 3 – press 3, if you want to talk with someone in our creative commerce department.
Step 4 – press 4, if you would like to talk to someone about our products.
Step 5 – press 5, if you would like to listen to a recording describing our services.
Step 6 – press 6, if you want to order some more of a book we have published.
Step 7 – press zero, if you want to talk to a specific individual and don't have the extension number.

I chose to press 4 and after it rang 8–10 times, I heard a very soft voice saying "I am Mary Jones and I am the lead author's representative at company ABC. I am presently talking with another very important author but your call is important to me, so please leave your phone number and I'll call you right back." She never did.

Eventually I gave up on trying to contact a representative from the organization and was able to find Mary Jones's email address. One day after I sent my email to her, I received a phone call explaining that she is very sorry she had not reacted to my previous contacts, but she has been very busy and hadn't had time to listen to any of the messages that were left on her telephone.

Well, that should've been enough to convince me not to do business with this organization, but she had a sweet voice and took time to explain in detail the services they provided. She convinced me that I did not know enough about self-publishing to do it myself and that their promotional staff had contacts that would get my book reviewed in many key publications, almost guaranteed that it would be in the top-20 books. To make up for some of the inconvenience, she agreed to reduce the processing costs by approximately 25%. We agreed, and I immediately sent them a copy of my manuscript. About two days later, I received a bill from them, but they forgot to deduct 25% that Mary Jones had agreed to. I got a hold of Mary Jones again and she assured me she'd take care of it, and within another two days I received a corrected bill indicating that they had already deducted the full amount from my credit card account and they would be forwarding to an organization the 25% reduction.

This left me wondering how they could do this because I had never given them my credit card number or authorization to take money out of the credit account. On closer study, I realized that the credit card number that they used was not one that I recognized. After a discussion with their accounting department, we realized that they had used someone else's credit card number. They wanted me to give them information related to one of my credit cards so they could process the billing correctly. Needless to say it was time for me to look for another publisher. I believe this publisher is a good publisher, but they just had bad after-sales service.

Process Grouping 10: After-Sales Services Activity Block Diagram

The following Process Grouping 10. After-Sales Service Activity Block Diagram provides a view of some of the more critical documentation required to support a project for any new innovative products (see Figure 8.2). The following is more detailed descriptions of the activities that make up the documentation.

Inputs from Process Grouping 9 and the Knowledge Management system

10.1. Perform a simulated customer process walk-through to identify and document key customer touch points.

10.2. Train all after-sales personnel in customer relationship management and the outputs/technologies they will be addressing. Remember they are also the customer's only contact with the organization and they have to really be able to provide outstanding service and impressions for the

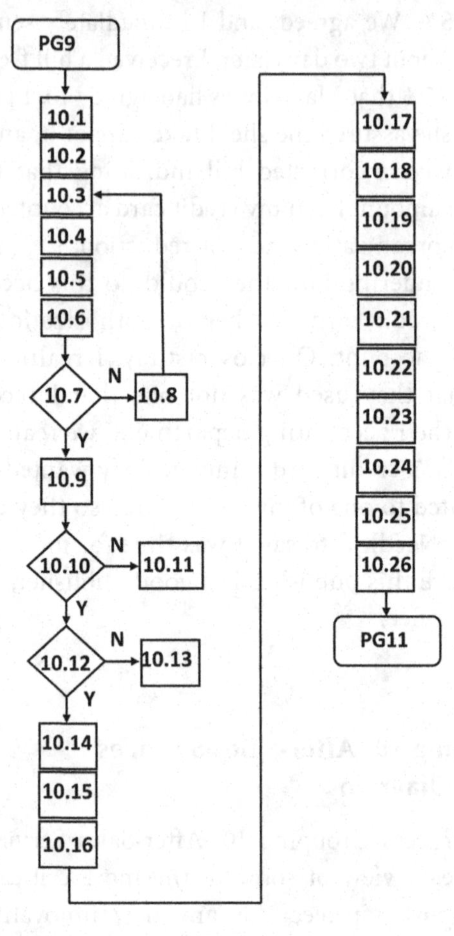

FIGURE 8.2
Process Grouping 10. After-Sales Service Activity Block Diagram

organization to survive. Where possible, provide the individual with a script that they can use as guidance when they talk with the customer. Identify activities that should be performed by only certified employees.

10.3. Review the knowledge management system to determine previous problems and corrective actions.

10.4. Collect current procedures related to after-sales services, and prepare procedures for activities where procedures are not documented but should be.

10.5. Perform a risk analysis of each customer's touch point.

10.6. Compare a list of all customer checkpoints to related documentation.

10.7. Does the documentation adequately mitigate risks related to the checkpoint?

10.8. If the risk is not adequately contained, develop an improved process.

10.9. Conduct a secret shopper analysis of major competition.

10.10. Do the documented mitigation plans for each of the customer touch points provide the organization with a competitive advantage?

10.11. If not, rewrite them.

10.12. Is the repair center located so that customer's downtime is minimized?

10.13. If not, relocate if practical.

10.14. Secure software that will allow the customer to pinpoint the status and location of his or her order.

10.15. Document agreed-to entity key replacement parts' arrival dates at the customer's location.

10.16. Set up a field failure analysis center(s).

10.17. Set up call center data collection and measurement system. Typical measurements for a call center include

- brief sponsor's time to call,
- number of calls satisfied the first time,
- customer satisfaction level with call center,
- length of call time, and
- performance by individual call center personnel.

10.18. Give the person calling the option of talking through the computer or to a real live person.

10.19. Give the call center personnel enough authority to negotiate a win–win situation between the organization and the unhappy customer.

10.20. Compare call center activities to best practices. Evaluate differences and take appropriate corrective action.

10.21. Ensure that call center personnel are getting breaks that are frequent enough and long enough to relieve the stress of their jobs (e.g., every 90 minutes the call-center personnel should get a 15-minute break). Breaks should be staggered so that people are available to service the callers without keeping them waiting for more than 10 minutes.

10.22. Conduct periodic customer-satisfaction surveys. Any item rated as "below outstanding" should be considered for continuous improvement.

10.23. Compare call center activities to best practices and update as appropriate.

10.24. Feed information into the knowledge management system that may help with design and layout of future entities.

10.25. Initiate request for corrective action for repetitive problems.

10.26. Output goes to the Knowledge Management System and Process Grouping.

Forward to PG 11.

After-sales service includes individuals who man the control center, handle customer complaints and customer questions, and provide a direct line of interface between the organization and its clientele. Another key part of after-sales service is a repair center.

PROCESS GROUPING 11: PERFORMANCE ANALYSIS

During this activity, data is collected to determine if the actual results meet or exceed the commitments in the business plan analysis stage. A post postmortem should be conducted before the project is closed out. This will provide input, both positive and negative, into the knowledge management system to help optimize future projects. Usually based upon this analysis, individuals doing outstanding work are rewarded and recognized.

It should not be difficult for the team to determine how they are going to measure the success of the project. Specific commitments were well documented in Tollgates II, III, and IV. These commitments often are based upon the entity's performance parameters, the cost to make the changes, the amount of resources consumed by the project, and the impact the new entity has on the market. The ideal situation is to evaluate these key parameters when you can run the new and previous entity at the same time without the impacted individuals knowing which entity they were working with. The medical field provides us with a very good example of this type of analysis when they bring together a group of individuals, half of who take the old cure and the other half takes the new cure. It is well understood that the psychological impact of the additional attention that the process is given when it is considered for change can result in major improvement in efficiency and effectiveness and adaptability that would have the same impact on both the new and old entities.

Another problem that the team faces is the learning curve for the new entity. The information gathered during opportunity identification and opportunity development was from the established line that is down on the flat of the learning curve, while the new item may be just on the improvement slope of the learning curve. As a result, timing and experience level of the production and sales can have a major impact upon the final results and conclusions.

TOLLGATE V: PROJECT EVALUATION

Typically, the project management team has completed their major activities when they successfully pass Tollgate IV, and both the team members and management are anxious to reapply their skills to another assignment. This often results in the performance analysis being completed before sufficient data is available. This is particularly true of customer-related and reliability-related measurements. Typical project goals would read like this: "The relation of zzk2 processor will increase our share of the market by 30%." The question is, "How do you possibly measure this when you want to complete Tollgate V within three months of the first time the customer product has been delivered to a customer?" In truth, we are continuously pressured by management to eliminate the project management team right after the first-customer ship and the knowledge management system has been updated to reflect things learned during the PIC.

Typical Improvement Methodology Results

The improvement approaches, when implemented correctly, can have a positive impact on the organization as noted here:

- Consume less resources
- Increase storage capacity
- Increase productivity
- Increase ability to handle variation
- Increase customer satisfaction
- Increase number of employees
- Increase profits

- Increase market share.
- Increase return on investment.
- Increase value added per employee
- Increase stock prices
- Improve morale
- Improve quality
- Increase processing speed
- Improve customer satisfaction
- Improve competitive position
- Improve productivity
- Improve adaptability
- Improve reliability
- Improve maintainability
- Improve safety
- Decrease size of entity
- Decrease poor quality cost
- Decrease waste
- Decrease overhand
- Decrease inventory
- Decrease or eliminate layoffs
- Percent of revenue produced by the entity
- Reduce weight
- Reduce process cycling time
- Reduce process inventory
- Reduce costs
- The breakeven point

Tollgate V is a comparison of the actual performance of the project and its assigned entity that is frequently used as the closeout of the project management activity when the responsibility for continuing smooth operation is transferred to production and sales.

Top Five Positive/Negative Innovation Change Impacts

In order to understand the complexity of trying to satisfy all stakeholders, we need to understand each stakeholder's priorities. The following tables list the six stakeholders and their top-five improvement priorities and their top-five negative change impacts.

- List 1.1 – Investors Measure of the Improvement in Priority Order
  Return on investment
  Stock prices
  Return on assets
  Market share
  Successful new products
- List 1.2 – Investors Measure of Negative Change Impacts in Priority Order
  Reduce stock prices
  Reduce dividends
  Lower profit levels
  Reduce market share
  Failure of new products
- List 2.1– Management Measure of the Improvement in Priority Order
  Return on assets
  Value added per employee
  Stock prices
  Market share
  Reduced operating expenses
- List 2.2 – Management Measure of Negative Change Impacts in Priority Order
  Increased operating costs
  Reduce market share
  Lower customer satisfaction levels
  Failure of new products
  Longer cycle times
- List 3.1– External Customer Measure of the Improvement in Priority Order
  Reduce costs
  New or expanded capabilities
  Improved performance/reliability
  Ease to use
  Improved responsiveness
- List 3.2 – External Customer Measure of Negative Change Impacts in Priority Order
  Increase purchase costs
  Decreased reliability
  Fewer capabilities than competition

Poor customer service
Increased difficulty to use

- List 4.1 – Their Employees Measure of the Improvement in Priority Order
  Increase job security
  Increased compensation
  Improved personal growth potential
  Improve job satisfaction
  Improve management
- List 4.2– Employee Measure of Negative Change Impacts in Priority Order
  Layoffs
  Decreased benefits
  Salaries not keeping pace with cost-of-living
  Poor management
  Decreased skills required to do the job (boarding work)
- List 5.1 – Suppliers Measure of the Improvement in Priority Order
  Increased return on investment (supplier)
  Improved communications/fewer interfaces
  Simplified requirements/fewer changes
  Long-term contracts
  Longer cycle times
- List 5.2 – Suppliers Measure of Negative Change Impacts in Priority Order
  Loss of contract
  Shorter order cycles
  Increased competition
  Imposing new standards
  Longer accounts payable cycle times
- List 6.1– Community Measure of the Improvement in Priority Order
  Increasing employment of people
  Increased tax base
  Reduce pollution
  Support of community activities
  Safety for employees
- List 6.2 – Community Measure of Negative Change Impacts in Priority Order
  Moving work overseas

Decreasing the number of employees
Decreased facility resulting in lower taxes
Unsafe working conditions
Increased pollution of the environment (increase in toxic gases and materials)

In evaluating the positive and negative impacts that a change initiative can generate in the organization's stakeholders, we use the following approach. For each of the six stakeholders, we evaluate the five negative and five positive impacts' rating based upon a scale of 1 to 5. A rating of 1 indicates that you are in strong disagreement with the statement, and a reading of 5 indicates you strongly are in agreement that the statement applies to your organization. We then sum up all of the positive and negative measurements, keeping them separate from each other. Next, we subtract the negative change measurements from the positive change measurements. The higher this number is, the better. A score of 25 is the best possible score, and anything under 15 indicates a major change is required in order to service your stakeholders properly.

The ideal improvement process would improve the organization's performance in all the stakeholders' priorities issues with lesser impact upon the negative change impacts. In these tables, the most frequent impact was noted, but sometimes one methodology can have more than one impact, depending upon the circumstances. For example, Total Quality Management (TQM) can have a positive or negative impact on job security. If improving the product increases the organization's market share resulting in an increased workload, job security is improved. But if TQM results in waste reduction, thereby improving productivity, but does not increase market share to the point that it offsets the productivity gain, employees can be laid off. This results in a negative impact upon job security. It's easy to see that if an organization is a nuclear power plant, safety would be the number one priority for the management, community, and the employees.

Process Grouping 11: Performance Analysis Activity Block Diagram

The following Process Grouping 11. Performance Analysis Activity Block Diagram provides a view of some of the more critical documentation required to support a project for any new innovative product. The following

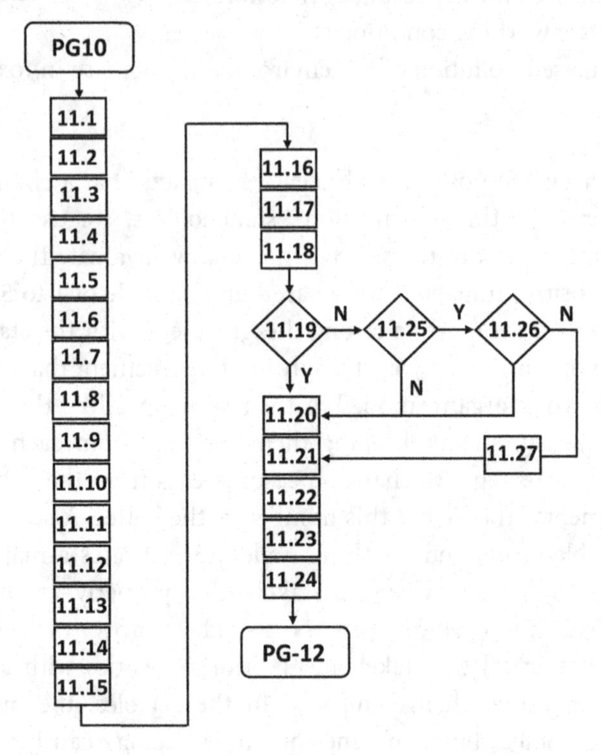

FIGURE 8.3
Process Grouping 11. Performance Analysis and Tollgate V Project Evaluation Activity Block Diagram

is more detailed descriptions of the individual activities that make up the Activity Block Diagram (see Figure 8.3).

Input to Process Grouping 11 outputs from Process Groupings 1 to 4 and the knowledge management system entries related to the entity assigned to the project team.

11.1. Review the Business Case Analysis, listing all committed improvement parameters and related resources.

11.2. Review the opportunity identification statement that was output from Process Grouping 1.

11.3. Make a list of all of the hard and soft impacts the project would have on the organization along with where the commitment was made and the date that it was made.

11.4. Use the commitments made in Tollgate III as the standard requirement for the project. Tollgate III was used as the standard because it is the point where the project became part of the organization's portfolio of projects.

11.5. Develop a test plan to collect information related to the committed measurements. Select a sample size that is large enough to get statistically sound results and report confidence limits for the results.

11.6. Train impacted employees to collect data in a manner as close to the way the original data was collected.

11.7. Observe initial collection at each collection point. Then pick up data daily and analyze to be sure that it is usable data.

11.8. Compare the projected improvement impacts as projected at Tollgate III.

11.9. Calculate the change impact on the investor's priorities.

11.10. Calculate the change impact on the management's priorities.

11.11. Calculate the change impact on the customer's priorities.

11.12. Calculate the change impact on the employee's priorities.

11.13. Calculate how the change impact on the supplier's priorities.

11.14. Calculate the change impact on the community's priorities.

11.15. Check to ensure that the control system is in place that will ensure the gains made as a result of the project/program is not degraded over time.

11.16. Prepare a final report that includes information related to each committed change, the confidence level related to each data point.

11.17. Include in the report the team's observations, suggestions, conclusions, and recommendations. They should also include a section discussing the lessons learned, the problems faced, and the additional knowledge that was acquired as a result of participating in the project.

11.18. Present the report to the executive team and get permission to reassign the team members.

11.19. Did the project meet requirements?

11.20. Yes – Document result.

11.21. Continue production.

11.22. Document results of Tollgate V. Project Evaluation.

11.23. Update the individual personnel file reflecting the contributions they made to the project.

11.24. Update the knowledge management system.
11.25. No – Is additional change required?
11.26. Can the situation be corrected?
11.27. Change requirements to agree with product performance capabilities.

Output – Dissolved project team recognizing and thanking each of the team members for their contributions.

Summary

During this activity, data is collected to determine if the actual results meet or exceed the committed improvements as defined in Phases I and II. This analysis often is difficult because the new item is at the beginning of its learning curve, keeping it from reflecting its true savings. An offsetting factor is the fact that the additional attention a new entity gets will bring about improvement that could have been obtained in the old entity if equal attention was given to its importance. Software packages frequently do not result in the promised improvement particularly when it's being compared to the competition. The reason is that the competition is often installing the same software package, and as a result, there is little or no competitive advantage induced into the organization when the new software is brought online. The result is installing the new software did not create a positive competitive position for the organization. Yet it does not mean that there was not improvement.

The other side of the coin is if you did not install the software packages, the organization would be at a considerable disadvantage in comparison to its competition that did install software that improves their efficiency and/or effectiveness. It just means that organizations today have to continuously invest in improvements, often generating a negative impact upon the bottom line as the costs of the software and its insulation detracts from the bottom line. A post postmortem should be conducted before the project is closed out. This will provide input both positive and negative into the knowledge management system to help optimize future projects. Usually based upon this analysis, individuals doing outstanding work are rewarded or recognized.

PROCESS GROUPING 12: TRANSFORMATION

Usually the project team is disbanded after the Process Grouping 11 is completed, but that's only the beginning of the project story. The real test of the project occurs over the next year or two when the approaches are often reset to the original habit patterns. For successful innovative projects, changes have to become part of the organization's culture and habit patterns. This is where the real impact of the project is evaluated.

Most of the innovative measurements of success are related to products; unfortunately that is a poor measure of the real impact. The real success is measured by the changing behavioral patterns of your employees and managers. It is for that very reason that when you first start an improvement initiative, you always define how you want employee's and management's behaviors to change, for that's what changes the culture within your organization.

The key commitment document for a project is the project plan as approved during Process Grouping 5. Business Case Analysis. Part of the process plan at this point in time is a mission statement that defines the length of time that would be required to complete the project, the total cost of the project, the impact the project will have on the organization's performance and behavioral patterns. After studying and thoroughly understanding these process commitments, the team should review the Tollgate V. Project Evaluation final report to understand how implementing the project recommendations have impacted the organization's performance initially. The team should then verify that the changes that were implemented as part of the project are still being used and considered value added by the individuals involved in the changed activities/approaches and impacted areas.

The team will also verify that at that point in time, the actual savings were in keeping with the projected estimates. It is very important that the team also validate that the savings went directly to the bottom line of the key performance measurements rather than being redeployed to do something that wasn't being done before because it wasn't justified. All too often savings that save an individual 10 minutes a day amount to longer coffee breaks and as result cannot be counted as value-added activities. The savings can only be realized when the individuals who were performing the eliminated activities have been reassigned and are performing a job that generates more real value added than the old assignment did. Dismissing the assignment of people who counted as value-added results is one of the primary reasons much of the improvement savings never reach the bottom line.

The cycle time reductions that don't result in increased sales have little or no impact on the bottom line. The only value as a result of the cycle time reduction is the rental space savings reduction and the interest rate on the value of the entity related to the decreased cycle time. Often reduced cycle time that results in lower inventories has a negative rather than a positive impact. This results in waste, like sending entities by overnight mail that could've been sent by regular mail.

How many times have you been put on hold for hours waiting to talk to someone in the suppliers' service area, when their customer representatives are all too busy on the phone. Often these long delays are caused by government offices that are not-for-profit, and as a result, they are assuring that my company will be not-for-profit. How many of you have lately been dissatisfied with one or more of your suppliers because you cannot get someone to talk to you because everything is computerized? How many of you have needed information from a supplier and was connected to a computer answering equipment that gives you a number of options to dial? Once you dial in, you are given a number of other options to select from, choose option number three only to be disconnected. How many of you have a better customer satisfaction index today than you had two years ago? How many of you have email messages on Sunday informing you that you have a zoom meeting on Monday at 6 AM and then have people unhappy with you because you don't make the meeting? Why is it I used to have a job where I worked only 12 hours a day, 6 days a week? Now to make a living, I have to work 24 hours a day, 7 days a week and my wife also has to work and take care of our children in the home. The measured value added has the amount of time we free up to spend with our family.

This evaluation focuses on the sustainability and acceptance level of the innovative entity or change resulting from the project being assessed. The first task is to define, acquire, and understand the documents that define the improvement opportunity, its analysis, and the action that was taken to take advantage of the opportunity. The following is a list of some of the information sources that may be required as input into the analysis.

- Appropriate engineering specifications
- The project plans
- Knowledge management, warehouse

- The Primary Performance Measurements
- Organizational change management plan
- Entity performance reports (quality, reliability, delivery, and customer complaint)
- Amendments to the project plan after initial approval
- Tollgate II final report
- Tollgate III final report
- Tollgate V final report
- Project financial final summary
- Implementation cost of Tollgate III recommendations
- Output production costs since start of project
- Revenue generated as a result of project implementation
- Customers' opinion surveys related to project impact on them
- Budget expenditures and decreases related to the project
- Employee turnover rate
- Postmortem evaluation of the project by the project team
- Warranty summary report

For the purpose of the Activity Block Diagram that is presented here, we will include all of these under the heading "Tollgate V+ other inputs."

Process Grouping 12: Transformation Activity Block Diagram

I believe we all are aware that there is a big difference between what we say we are going to do and what we actually do. "I'll get the report out Monday by 10 o'clock." You finally got it out late Tuesday night. There's a big difference between what we say we are doing and what we do every day. "I brush my teeth after every meal. Never less than twice a day." In reality, there are many days each year that I brush my teeth only once a day and even on some very rare occasions, I don't have time to brush my teeth all day.

I am writing a book titled *Boy with Three Lives*. It is more than half written, and last summer I told my son I would have it completed before Christmas. Here it is almost a year later and I haven't had time to even look at it yet. In reality, I doubt if I'll have time to do anything with it during the rest of this year.

We find that for many project plans that we ran across, the author has all sorts of good intentions, but other things get in the way and priorities change. Our CEO is mad about something that pushes my priorities way down on my priority list, and I often don't get back to doing everything I wanted to do.

For these and many other reasons, it is not prudent to assume that everything that was committed was completed.

Here is a list of typical activities that would be addressed in conducting Process Grouping 12. Transformation Analysis. The following Process Grouping. Performance Analysis Activity Block Diagram provides a view of some of the more critical documentation required to support a project for any new innovative product (see Figure 8.4).

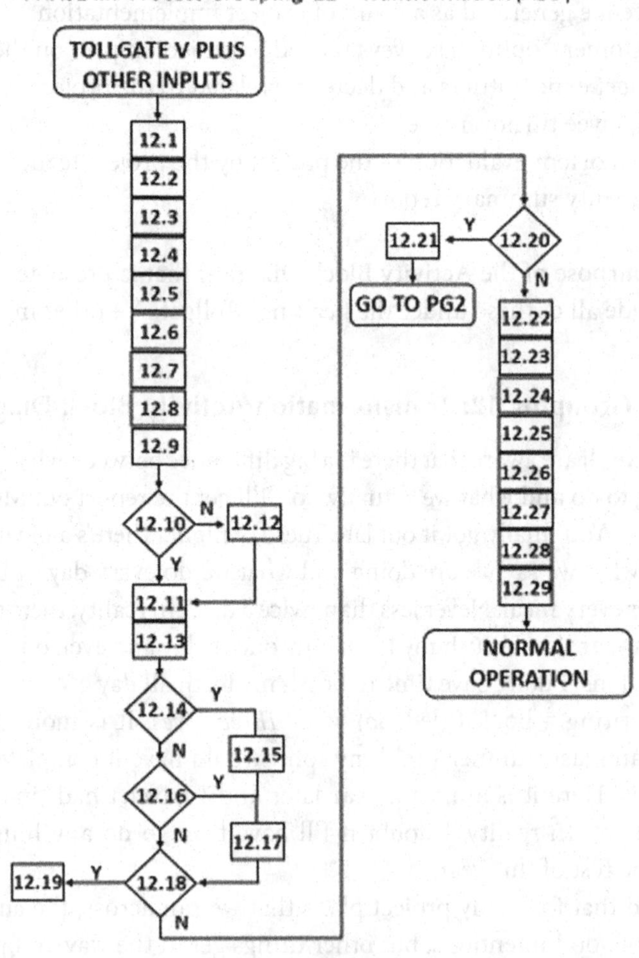

FIGURE 8.4
PHASE III: Process Grouping 12. Transformation Activity Block Diagram

Input – "Tollgate V+ other inputs."

12.1. Assign a team to conduct Process Grouping 12. Transformation Analysis.

12.2. Acquaint the team with the activities that took place to take advantage of the improvement opportunity.

12.3. Get a copy of Tollgate III final report, and make a list of projected hard and soft improvement results.

12.4. Get a copy of Tollgate V final report, and compare actual hard and soft improvement results to projected improvement results.

12.5. Acquaint the team with the entity production cycle. Conduct a process walk-through acquainting the entire team with the project/ program changes and results. Develop checklist to collect information from individuals involved in the activities. During the conversation, be sure to ask if there is anything that should be done to improve the activities he or she was involved in. I always like the question, "What would you do differently if you were the manager of this area?"

12.6. Collect total cost spent to take advantage of the opportunity that was over and above the cost of the entity itself.

12.7. Collect the product costs, including salaries to produce, the materials, and overhead.

12.8. Get the accounts receivable based upon the entities that were delivered to the customer.

12.9. Define breakeven point.

12.10. Compare actual costs and timing for the project's cost and completion dae as projected in Process Grouping 6. Business analysis. Based upon the information in the knowledge warehouse and information you have now, would you still have approved the project during the Business Case Analysis?

12.11. Yes – Continue the analysis.

12.12. No – Ok to go to continue the analysis, but assign a group to upgrade the early screening methodologies to improve decision-making.

12.13. Review field performance.

12.14. Update sales and marketing plan.

12.15. Is there a safety problem?

12.16. Yes – Design a permanent fix to the safety problem.

12.17. Does the entity need to be more reliable?

12.18. Yes – Analyze root cause of failure and implement corrective action in the process.

12.19. Is an entity customer/consumer recall required?

12.20. Yes – Repair or replace entities already delivered to customers/ consumers.

12.21. Does the entity need additional features to attract additional customers or to be competitive?

12.22. Yes – Go to Process Grouping 2. Opportunity Development.

12.23. Analyze the opportunities that the project took advantage of and implemented improvements that were value added to determine which activities should be applied to the general operations and future organizational outputs.

12.24. Update general operating procedures, control systems, and knowledge warehouse, making the activities identified in 12.22 part of the organization's basic systems.

12.25. Distribute and train individuals in the use of the updated general operating procedures and control systems.

12.26. Now define who the organization's stakeholders are. Typical stakeholders would be as mentioned here:

- Customers
- Consumers
- Investors
- Management
- Employees
- Employees families
- Suppliers
- Interested parties
- Government

12.27. For each of these stakeholders, define how you think the stakeholder would feel about the project and its impact upon him/her. Select the statement that best reflects the way each stakeholder would view the action taken:

- Very important /absolutely the right thing to do
- Important/it was worthwhile doing
- Minor/minor positive change but may not be worth the investment
- No reaction/a waste of effort from my standpoint
- Negative reaction/it was a waste of effort
- Very negative reaction/another example of someone taking care of themselves and nothing about me
- Extreme negative impact upon my lifestyle/was people like me out of a job and greatly downgrades our living standards

12.28. Estimate how the project value added and negative value content of the project will impact each of the Primary Performance Measurements. (The following is the suggestions on ways that could help define the impact of an individual project on each of the Primary Performance Measurements.)

- Measurable bottom-line impact
- Slight improvement
- No impact
- Slightly negative impact
- Measurable impact

12.29. Now the executive team defines if the decision to invest in this opportunity was a good or bad decision by rating it (see the following typical categories).

- A bad decision
- Not the best way to use the resources
- A good decision but could've been better
- A good decision
- Absolutely the correct decision

Document the activities that took place during the transformation analysis and update the knowledge warehouse.

Output – Project activities turned over to normal operations.

SUMMARY

The PIC is a continuous flow activity with a number of loops to take advantage of additional data that becomes available. Treat it like a process and it will behave well; treat it like a lot of little pieces and it will bite you every time. Like any process, it has to have a start and finish. The starting point for the PIC is the search for an opportunity to apply your creative powers to bring about a new and unique answer to a previously unanswered situation. It ends when output from the process delights the projected user.

I innovate today to eat tomorrow.

—H. James Harrington

9

Special Mention Tools and Methodologies

INTRODUCTION

There is a myriad of different tools and methodologies that have been developed to help an organization or individuals be creative/innovative. A working group was formed by the International Association of Innovation Professionals (IAOIP) to evaluate the tools and methodologies that were presently being used and determine which were the most useful. To accomplish this mission, the working group studied the literature that was currently available to define tools and methodologies that were presently proposed or being used. They also contacted numerous universities that were teaching classes and innovation and entrepreneurship to determine what tools and methodologies they were promoting. In addition, they contacted individual consultants who are providing advice and guidance to organizations in order to identify tools and methodologies that they were recommending. As a result of this research, more than 200 schools and methodologies were identified as being potential candidates for the innovative professional. They then made a survey where most of the 200 tools were analyzed, and giving a point score of 0 to 4, the target was to reduce the list down to 76 tools/methodologies. The tools and methodologies were then divided into three classifications: creative stools/methodologies, evolutionary or improvement tools/methodologies, and organizational, operational tools/methodologies.

Based upon this study, they divided the manuscripts up into the following three books published by Taylor and Francis group:

- *The Innovation Tools Handbook, Volume 1: Organizational and Operational Tools, Methods, and Techniques That Every Innovator Must Know* (2016)

DOI: 10.4324/b22993-9

- *The Innovation Tools Handbook, Volume 2: Evolutionary and Improvement Tools That Every Innovator Must Know* (2016)
- *The Innovation Tools Handbook, Volume 3: Creative Tools, Methods, and Techniques That Every Innovator Must Know* (2016)

A list of these tools and their classifications is included in Appendix B of this book.

The following are four of these tools, methods, and techniques that we want to highlight as you consider improving your organization's innovation.

AREA ACTIVITY ANALYSIS: A KEY 2020S TOOL PUT ON THE BACK BURNER

Area Activity Analysis (AAA), sometimes called "Department Activity Analysis," is the most effective approach developed today to establish the internal and external supplier/customer relationships. It also establishes Individual Performance Indicators (IPI) related to major activities that go on in all of the organization's natural work teams (NWTs). These individual performance indicators measure what each level of management and all employees do and what they are responsible for. This allows improvement efforts to be focused upon real problems and to take advantage of real opportunities that impact the total organization's performance. Using this approach, the total organization's performance typically increases at a rate of 10% to 20% per year.

Introduction

In the 1980s and 1990s, there was a major focus on setting up the internal and external customer–supplier relationships within an organization. Although the external customer was given priority, organizations began to realize that the best way to have external customer satisfaction was to have very satisfied internal customers. Many experiments proved that as we improve the processes that service our internal customers (our employees), we provided much better services for our external customers. This realization drove the

Business Process Improvement methodologies and the Quality Circles approaches. We started training everyone in simple problem solving. Total Quality Control (TQC) gave way to Total Quality Management (TQM). Elite special Six Sigma teams were trained and assigned to attack the big problems. Toyota's Lean Production Systems became the established standard of excellence as we moved into the 21st century. Quality improvement is now being measured by cost reduction savings, not external customer satisfaction. Six Sigma Black Belts are now held accountable for saving $1 million a year, not for improving external customer satisfaction by *X* percent.

It may be time for the quality professionals to stop focusing on cost reduction and directing their efforts at improving internal and external customer satisfaction. We need to get everyone in the organization to go beyond doing their best and start giving their all. We need to start to develop a family spirit within the total organization. It is time to help every employee realize that each of them plays a key role in the success of the organization. Every job is a value-added job and contributes to meeting the organization's goals. If it doesn't, we shouldn't be doing it. This is where a little used tool that was developed in the 1980s can come into play. That tool is called "Area Activity Analysis" (AAA). In my mind, this tool is more important and will have a bigger impact upon the total organization than Six Sigma and Lean combined. It is the tool that all organizations should use as a start of their continuous improvement activities.

- Definition: Area Activity Analysis is a methodology to establish agreed-to, understandable efficiency and effectiveness measurement systems and communication links throughout the organization. It is designed to define and set up all of the internal and external supplier/customer relationships (see Figure 9.1).

The methodology consists of seven phases. The first six phases of AAA process usually have fixed start- and end-points; only the last phase, Continuous Improvement, is a continuous process. The first six phases are normally treated as a project and they are included in the organization's strategic business plan. This by no means indicates that the measurements and associated requirements are developed and not updated on a regular basis.

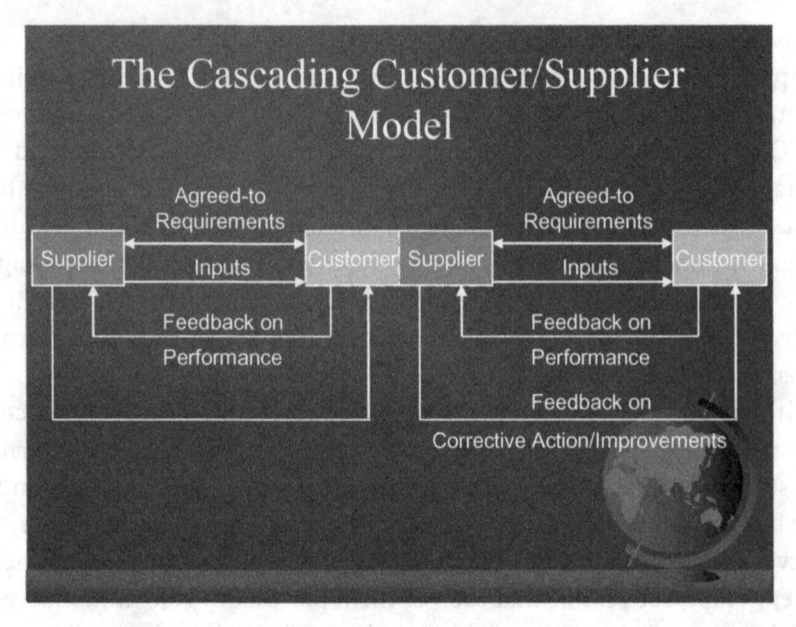

FIGURE 9.1
The Cascading Customer/Supplier Model

Benefits of AAA

AAA has many benefits to an organization. Some of them are:

- It aligns the mission statement throughout the organization from the president's office all the way down to first-line-foremen's group (Natural Work Team).
- It ensures that the individual Natural Work Teams (NWTs) are conducting activities that are in line with their mission statements.
- It provides a good understanding by all of the NWT members of what is going on within the NWT.
- It provides agreed-to, understandable output requirements as viewed by the internal and external customers.
- It establishes a performance-feedback system from the internal and external customers.
- It establishes agreed-to and documented work standards for all major processes within the NWT scope of work.
- It documents the major processes that go on within each NWT.
- It defines and documents the input efficiency and effectiveness requirements for all inputs to the NWT major processes.

- It sets up a visual and performance and problem tracking system for each NWT.

Definition: Natural Work Team (NWT) or Natural Work Group (NWG) is a group of people who are assigned to work together and who report to the same manager or supervisor. AAA projects are implemented by NWTs.

(Note: The COO and all the vice presidents who report to him or her make up an NWT just as the card assembly line foreperson and all his or her employees make up an NWT.)

Features of AAA

The features of a typical AAA methodology include:

- It develops aligned, interdependent mission statements for NWTs at all levels within the organization from the CEOs of NWT to the janitorial NWT.
- It defines the major processes that are going on in each NWT.
- It documents customer agreed-to, signed-off output requirements for the major processes that each NWT is involved in.
- It defines and documents the performance feedback loop for the internal and external customers for each output from the major NWT processes.
- It defines and documents signed-off work standards for the major processes for each NWT.
- It documents the processes that create the output for each NWT.
- It documents a set of supplier agreed-to input requirements for each major process for each NWT.

Background

The first AAA activities were started in IBM by H. James Harrington in the early part of the 1980s. It was called "Department Activity Analysis." In the late 1980s, Norm Howery worked with Harrington to develop a training manual that was made available to the general public through Harrington, Hurd, and Rieker consulting firm. In 1995, Ted Cocheu updated the training manual for Ernst & Young. The experience gained by consultants, such as David Farrell and Ken Lomax, in using AAA as a service to their clients provided a major contribution to the development and improvement of the methodology during the 1990s. In 1999, McGraw-Hill published *Area Activity Analysis – Aligning Work Activities and Measurements to Enhance Business Performance*, authored by H. James Harrington. Also in the same year a CD-ROM entitled "Area Activity Analysis" was published for general public sales.

AAA is a methodology used by IBM, Ernst & Young, and other firms since the late 1980s.

> *Few incentives are more powerful than membership in a small group engaged in a common task, sharing the risks of defeat and the potential rewards of victory.*
>
> —**Robert B. Reich,** *The Work of Nations: Preparing Ourselves for the 21st Century Capitalism* **(Vintage Books 1992)**

AAA Methodology

The AAA methodology has been divided into seven different phases to make it simple for the NWT to implement the concept. Each of these phases contains a set of steps that will progressively lead the NWT through the methodology. These seven phases are provided in Table 9.1.

We will briefly describe each of the seven phases of AAA. Implementing these seven phases will bring about a major improvement in the organization's measurement systems, increase understanding and cooperation, and lead to reduced cost and improved quality throughout the organization.

TABLE 9.1

Seven Phases of AAA

Phase	Number of Steps
Phase I – Preparation for AAA	5
Phase II – Develop Area Mission Statement	5
Phase III – Define Area Activities	8
Phase IV – Develop Customer Relationships	7
Phase V – Analyze the Activity's Efficiency	4
Phase VI – Develop Supplier Partnerships	6
Phase VII – Performance Improvement	8
Total:	43

Phase I: Preparation for AAA

AAA is most effective when it precedes other initiatives that are taking place such as continuous improvement, team problem solving, Total Quality Management, reengineering, or new IT systems. It is also best to implement the AAA methodology throughout the organization. This does not mean that it will not work if other improvement activities are underway or if it is only used by one area within the total organization. In the preparation phase, the good and bad considerations related to implementing AAA within an organization should be evaluated. A decision is made whether or not to use AAA within the organization. If the decision is made to use AAA, an implementation strategy is developed and approved by management. Phase I – Preparation for AAA is divided into the following five steps:

- Step 1. Analyze the Environment
- Step 2. Form an AAA Project Team
- Step 3. Define the Implementation Process
- Step 4. Involve Upper Management
- Step 5. Communicate AAA Objectives

Phase II: Develop Area Mission Statement

- Definition: Mission Statement – A mission statement is used to document the reasons for the organization's or area's existence. It is usually prepared prior to the organization or area being formed and is seldom changed. Normally, it is changed only when the organization or area decides to pursue a new or different set of activities.

For the AAA methodology, a mission statement is a short paragraph, no more than two or three sentences, that defines the area's role and its relationship with the rest of the organization or the external customer. An area's mission statement must reflect a part of the mission statement that the area reports to. A manager can't delegate responsibilities to a group who report to him or her that the manager is not responsible for (see Figure 9.2).

Every area should have a mission statement that defines why it was created. It is used to provide the area manager and the area employees with guidance related to the activities on which the area should expand its

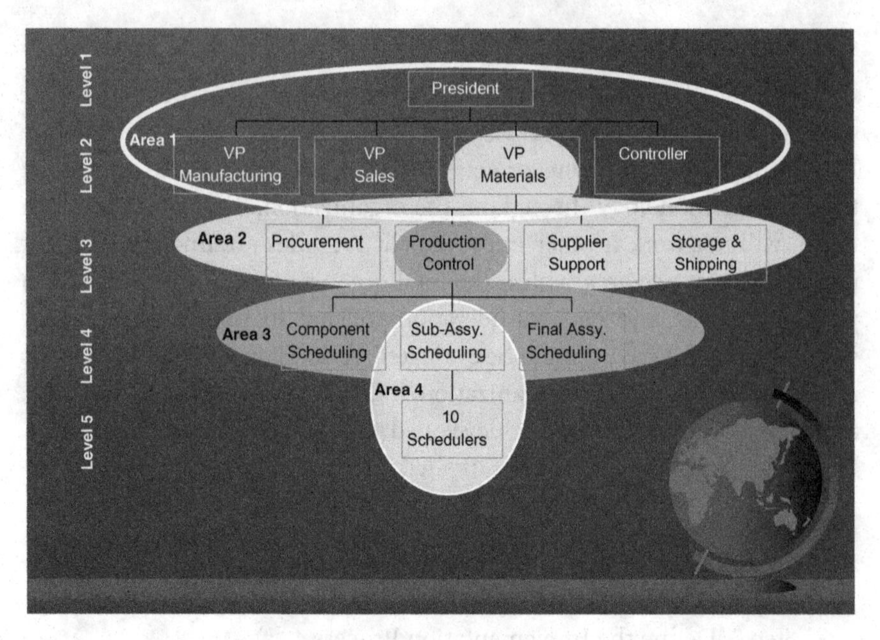

FIGURE 9.2
How AAA Worked in a Five-Level Organization

resources. The standard good business practice calls for the area's mission statement to be prepared before an area is formed. The mission statement should be reviewed each time there is a change to the organization's structure or a change to the area's responsibilities. It should also be reviewed about every four years, even if the organization's structure has remained unchanged, to be sure that the mission statement reflects the current activities that are performed within the area.

During Phase II, the NWT will review and update the area's mission statement. If a mission statement does not exist, the NWT will prepare a mission statement. In all cases, any change to the mission statement must be approved by upper management before it is finalized. Phase II – Develop Area Mission Statement is divided into the following six steps:

- Step 1. Obtain Present Mission Statement
- Step 2. Develop Preliminary Area Mission Statement – NWT Manager
- Step 3. Develop Preliminary Area Mission Statement – Each Employee
- Step 4. Develop a Consensus Draft Area Mission Statement

- Step 5. Finalize Area Mission Statement
- Step 6. Develop the Area's Service Policy

Phase III: Define Area Activities

During this phase, the NWT will define the activities that are performed within the area. For each major activity, the NWT will define the activity's output(s) and the customers that receive that output. Phase III – Define Area Activities is divided into the following eight steps:

- Step 1. Identify Major Activities – Each Individual
- Step 2. Combine into Broad Activity Categories
- Step 3. Develop Percentage of Time Expended
- Step 4. Identify Major Activities
- Step 5. Compare List to Area Mission Statement
- Step 6. Align Activities with Mission
- Step 7. Approval of the Area's Mission Statement and Major Activities
- Step 8. Assign Activity Champions

Phase IV: Develop Customer Relationships

During this phase, the NWT will meet with the internal and external customers who are receiving the outputs from the major activities conducted by the area to:

- Define the customer's requirements
- Develop how compliance to the requirements will be measured
- Define acceptable performance levels (performance standards)
- Define the customer feedback process

Phase IV – Develop Customer Relationships is divided into the following seven steps:

- Step 1. Select Critical Activity
- Step 2. Identify Customer(s) for Each Output
- Step 3. Define Customer Requirements
- Step 4. Define Measurements
- Step 5. Review with Customer

- Step 6. Define Feedback Procedure
- Step 7. Reconcile Customer Requirements with Mission and Activities

Phase V: Analyze the Activity's Efficiency

For each major activity, the NWT will define and understand the tasks that make up the activity. This is accomplished by analyzing each major activity for its value-added content. This can be accomplished by flowcharting or value-stream mapping the process and collecting efficiency information related to each task and the total process. The following is the typical information that would be collected:

- Cycle time
- Processing time
- Cost
- Rework rates
- Items processed per time period

Using this information, the NWT will establish efficiency measurements and performance targets for each efficiency measurement that is signed off by the person that the area reports to and the work standards group, if there is one. This work standard is often used to calculate the area's workload.

Phase V – Analyze the Activity's Efficiency is divided into the following six steps:

- Step 1. Define Efficiency Measurements
- Step 2. Understand the Current Activity
- Step 3. Define Data Reporting Systems
- Step 4. Define Performance Requirements
- Step 5. Approve Performance Standards
- Step 6. Establish a Performance Board

Phase VI: Develop Supplier Partnerships

Using the flowcharts generated in Phase V, the NWT identifies the supplier that provides input into the major activities. This phase uses the same approach discussed in Phase IV but turns the customer/supplier relationship around.

In this phase, the area is told to view itself in the role of the customer. The organizations that are providing the inputs to the NWT are called internal or external suppliers. The area (the customer) then meets with its suppliers to develop agreed-to input requirements. As a result of these negotiations, a supplier specification is prepared that includes a measurement system, performance standard, and feedback system for each input. This completes the customer/supplier chain for the area, as shown in Figure 9.1.

- Definition: Supplier – An organization which provides a product (input) to the customer (source ISO 8402).
- Definition: Internal Supplier – Areas within an organizational structure that provide input into other areas within the same organizational structure.
- Definition: External Supplier – Suppliers that are not part of the customer's organizational structure.

Phase VI – Develop Supplier Partnerships is divided into the following five steps:

- Step 1. Identify Supplier(s)
- Step 2. Define Requirements
- Step 3. Define Measurements and Performance Standards
- Step 4. Define Feedback Procedure
- Step 5. Obtain Supplier Agreement

Phase VII: Performance Improvement

This is the continuous improvement phase that should always come after a process has been defined and the related measurements are put in place. It may be a full TQM effort or just a redesign activity. It could be a minimum program of error correction and cost reduction or a full-blown total improvement management project.

During Phase VII, the NWT will enter into the problem solving and error prevention mode of operation. The measurement system should now be used to set challenge improvement targets. The NWT should now be trained to solve problems and take advantage of improvement opportunities. The individual efficiency and effectiveness measurements will be combined into a single performance index for the area. Typically, the area's key measurement graphs will be posted and updated regularly.

During Phase VII, management should show its appreciation to the NWTs and individuals who expended exceptional effort during the AAA project or who implemented major improvements.

Phase VII – Performance Improvement is divided into the following eight steps:

- Step 1. Set Up the Reporting Systems
- Step 2. Identify the Activities to Be Improved
- Step 3. Install Temporary Protection If Needed
- Step 4. Identify Measurements or Task to Be Improved
- Step 5. Find Best-Value Solutions
- Step 6. Implement Solutions
- Step 7. Remove Temporary Protection If Installed
- Step 8. Prevent Problem from Recurring

Summary

Too many of the improvement programs that I see being implemented today remind me of Don Quixote's quest. They are full of good intentions, but they are misdirected. They're limited to a few people's view of what is important and have little long-lasting substance. People are trained to use tools, but they are not given the time or the opportunity to use them. We are relying more and more on technology to interface with ourselves and our external customers, and less and less on our people. For example, people send text messages to the person sitting at the desk right beside them instead of talking to them. AAA provides a systematic way of helping each employee understand what is going on within the natural work team and how the natural work team and he or she fits into the bigger picture. It also establishes a direct connection to the people that get the output from each employee's activities, thereby building a sense of ownership within the individual.

In addition to having a very positive impact on all the employees, AAA also provides the organization with an effective way to measure and control the organization down to the individual level. The individual performance indicators become the key drivers in the continuous improvement process activities. Progress or the lack of progress at the activity level is continuously visible, so action can be taken immediately to keep it on track. Truly, AAA is a technique that all organizations should be using as the basis for building their team activities and their continuous improvement programs.

BENCHMARKING FOR BETTER INNOVATION

Benchmark to become better than the best.

—H. J. Harrington

Introduction

We shouldn't go another word forward until we have a common agreement about what the difference is between the benchmark and benchmarking.

- Benchmark (BMK) – Standard by which an item can be measured or judged. It can be anything that other things are using as a reference point.
- Benchmarking (BMKG) – A systematic way to identify, understand, and creatively evolve superior products, services, designs, equipment, processes, and practices to improve your organization's real performance.

When is benchmarking often used?

- Creation Phase – During this phase benchmarking is used to provide a benchmark for the criteria that a new product/process should be able to equal or beat in order to be competitive.
- Value Proposition Process – During this phase benchmarking is often used to understand and estimate the market potential for a new product or service.
- Performance Analysis Process – In this phase benchmarking is used to compare the performance and value added of a product or service compared to the competition. The information is used to help direct future product development cycles.

Knowing is halfway to getting there.

—H. James Harrington

A good benchmark is one that highlights an opportunity for improvement, not one that indicates you are the best.

—H. James Harrington

To be the best, you must:

- Know your strengths and limitations
- Recognize and understand the leading organizations
- Use the best processes
- Have the best design
- Never stop improving

There are four types of benchmarking:

- Product and Services
- Manufacturing Processes
- Business Processes
- Equipment

Darel Hall, past manager of Transportation and Planning at AT&T's MMS division, stated, "AT&T went into benchmarking for the right reason: to improve our business and culture."

A key to be successful at this is to have meaningful measurements and challenging targets. You need to understand how and why other organizations perform as they do. In addition, you need to close negative gaps between best practices and the performance of its organization's processes. All too often the management team does not know the level of performance that is possible for the organization to reach. They don't believe that major improvements are possible. They don't know how to bring about major breakthrough improvements in their key performance indicators (KPIs); the answer to this limitation is benchmarking.

Figure 9.3 shows the 10 steps in benchmarking that lead you to "best of breed." Benchmarking is a never-ending discovery and learning experience that identifies and validates the best items in order to integrate their best features into the organization's operations/products to increase their effectiveness, efficiency, and adaptability. As a result, they are able to maximize their value-added contributions to the organization. Benchmarking compares the organization and its parts to the best organization regardless

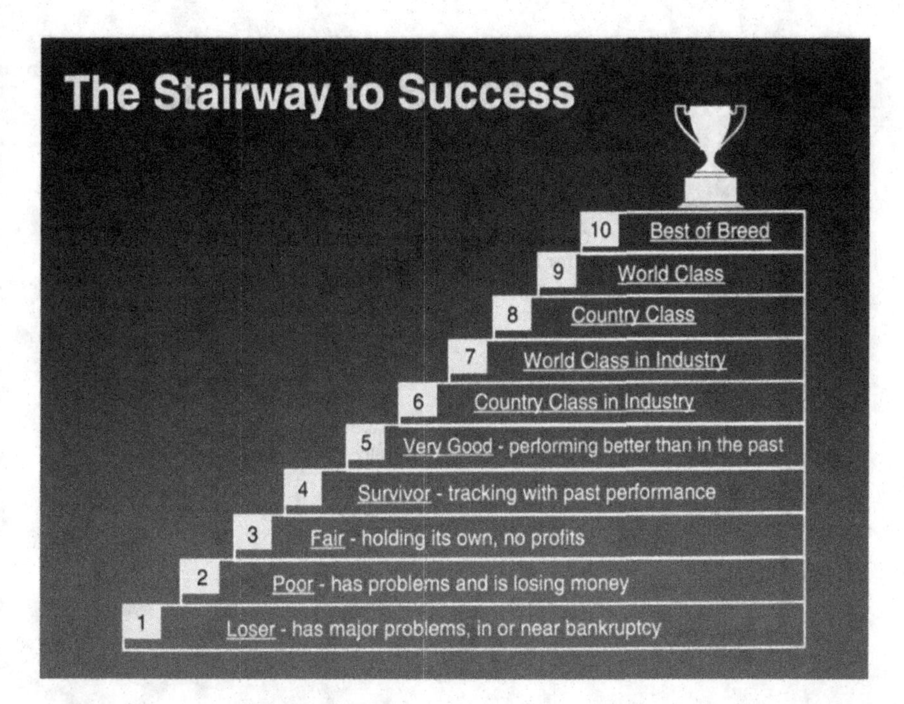

FIGURE 9.3
The Stairway to Success

of the industry or country. It helps the organization learn from others' experiences and shows the organization how it is performing compared to the best. Benchmarking is the antidote to self-imposed mediocrity.

Benchmarking is not a new methodology. Back in the 4th century B.C., Sun Tzu, author of *The Art of War*, wrote, "If you know your enemy and know yourself, the victory will not stand in doubt." Indeed throughout the ages, mankind has evaluated the strengths and weaknesses of others.

There are following five types of benchmarking processes:

- Internal Benchmarking
- External Competitive Benchmarking
- External Industrial (Comparable) Benchmarking
- External Generic (Trans Industry) Benchmarking
- Combined Internal and External Benchmarking

Figure 9.4 shows the differences in time it takes to do the benchmarking and the results that the approach gets on average.

Comparison of the Different Benchmarking Types

Benchmarking Type	Cycle Time for FSS	Benchmarking Partners	Results
Internal	3-4 Months	Within the Organization	Major Improvements
External Competitive	6-12 Months	None	Better than the Competition
External Industry	10-14 Months	Same Industry	Creative Breakthrough
Combined Internal & External	12-24 Months	All Industries Worldwide	Best-in-class
External Generic	12-24 Months	All Industries Worldwide	Changes the Rules

FIGURE 9.4
The Comparisons of the Different Types of Benchmarking

> *If AT&T had not been into quality, I'm not sure we could have pulled off benchmarking because of the culture it needs.*
>
> **—Edward Trucy, vice president, AT&T**

Benchmarking is a systematic process that has some specific ground rules that need to be understood and followed.

- Deciding what will be benchmark
- Defining the items to compare
- Developing measurements
- Defining sites and external organizations to benchmark
- Collecting and analyzing data
- Determining the gaps
- Developing future – state solutions
- Implementing the future – state solutions
- Measuring the impact

- Updating the benchmarking effort

The objective of benchmarking is to define places where you can improve, not to show how good you are.

—H. James Harrington

Table 9.2 provides a more detailed breakdown of the five phases and 20 activities of the Benchmarking Process.

Many people think that benchmarking is simply going out and studying other companies, but you must do a lot of homework before you go out.

—Tom Carter, past VP of ALCOA

TABLE 9.2

Detailed Breakdown of the Five Phases and 20 Activities of the Benchmarking Process

Benchmarking Phase	Related Activities
Phase I	1. Identify what to benchmark
Planning the Benchmarking Process and Characterization of the Item(s)	2. Obtain top management support
	3. Develop the measurement support
	4. Develop the data collection plan
	5. Review the plans with location experts
	6. Characterize the benchmark item
Phase II	7. Collect and analyze internal published information
Internal Data Collection and Analysis	8. Select potential internal benchmarking sites
	9. Collect internal original research information
	10. Conduct interviews and surveys
	11. Form an internal benchmarking committee
	12. Conduct internal site visits
Phase III	13. Collect external published information
External Data Collection and Analysis	14. Collect external original research information
Phase IV	15. Identify corrective actions
Improvement of the Item's Performance	16. Develop an implementation plan
	17. Gain top management approval of the future-state solution
	18. Implement the future-state solution and measure its impact
Phase V	19. Maintain the benchmarking database
Continuous Improvement	20. Implement continuous performance improvement

Benchmarking Cycle Time

The following provides insight into the time requirement in a sequence of activities required to conduct a benchmarking study with an external partner.

The Benchmarking Cycle

- Define two processes we will benchmark – April 25.
- Assign the benchmarking team – April 25.
- Define what and how they will be measured – April 30.
- Collect the internal data – May 15.
- Collect data from public domain sources – May 20.
- Analyze data to define three potential benchmarking partners – June 1.
- Contact potential benchmarking customers and get their agreement to be a benchmarking partner – June 20.
- Exchange information with benchmarking partners – July 10.
- Analyze data to determine if site visits are necessary – July 15.
- Conduct site visit – August 15.
- Distribute benchmarking findings and recommendations report – August 20.
- Implement action as is appropriate based upon management direction – August 25.

Evaluating Competitive Products

Most books and benchmarking classes completely overlook the techniques used to evaluate our competitors' products and services. This is a major error as it is one of the most movable benchmarking activities an organization can undertake.

- Services can be benchmarked through a technique called competitive shopping, sometimes called mystery shoppers.
- Products can be evaluated using a technique called reverse engineering.

Both of these techniques can be conducted only involving the organization's personnel without requiring the approval or cooperation of an outside organization.

Benchmarking Code of Conduct

Many people are worried about the ethics related to the benchmarking process. The following is a typical example of the code of ethics that all organizations should subscribe to before they start benchmarking.

Benchmarking Code of Conduct

Introduction

Benchmarking – the process of identifying and learning from good practices in other organizations – is a powerful tool in the quest for continuous improvement and performance breakthroughs. Adherence to this code will contribute to efficient, effective, and ethical benchmarking.

1. **Principle of Preparation**
   - 1.1. Demonstrate commitment to the efficiency and effectiveness of benchmarking by being prepared prior to making an initial benchmarking contact.
   - 1.2. Make the most of your benchmarking partner's time by being fully prepared for each exchange.
   - 1.3. Help your benchmarking partners prepare by providing them with a questionnaire and agenda prior to benchmarking visits.
   - 1.4. Before any benchmarking contacts, especially the sending of questionnaires, seek legal advice.
2. **Principle of Contact**
   - 2.1. Respect the corporate culture of partner organizations and work within mutually agreed procedures.
   - 2.2. Use benchmarking contacts designated by the partner organization if that is its preferred procedure.
   - 2.3. Agree with the designated benchmarking contact how communication or responsibility is to be delegated in the course of the benchmarking exercise. Check mutual understanding.
   - 2.4. Obtain an individual's permission before providing his or her name in response to a contact request.
   - 2.5. Avoid communicating a contact's name in an open forum without the contact's prior permission.
3. **Principle of Exchange**
   - 3.1. Be willing to provide the same type and level of information that you request from your benchmarking partner, provided that the principle of legality is observed.

3.2. Communicate fully and early in the relationship to clarify expectations, avoid misunderstandings, and establish mutual interest in the benchmarking exchange.
3.3. Be honest, complete, and timely with information submitted.

4. **Principle of Confidentiality**
4.1. Treat benchmarking findings as confidential to the individuals and organizations involved. Such information must not be communicated to third parties without the prior consent of the benchmarking partner who shared the information. When seeking for prior consent, make sure that you specify clearly what information is to be shared, and with whom.
4.2. An organization's participation in a study is confidential and should not be communicated externally without their prior permission.

5. **Principle of Use**
5.1. Use information obtained through benchmarking only for purposes stated to and agreed with the benchmarking partner.
5.2. The use of communication of a benchmarking partner's name with the data obtained or the practices observed requires the prior permission of that partner.
5.3. Contact lists or other contact information provided by benchmarking networks in any form may not be used for purposes other than benchmarking.

6. **Principle of Legality**
6.1. Take legal advice before launching any activity.
6.2. Avoid discussions or actions that could lead to or imply an interest in restraint of trade, market, or customer allocation schemes; price fixing; dealing arrangements; bid rigging or bribery. Do not discuss costs with competitors if costs are an element of pricing. Do not exchange forecasts or other information about future commercial intentions.
6.3. Refrain from the acquisition of information by any means that could be interpreted as improper, including the breach, or inducement of a breach, of any duty to maintain confidentiality.
6.4. Do not discuss, disclose, or use any confidential information that may have been obtained through improper means, or that was disclosed by another in violation of a duty to maintain confidentiality.

6.5. Do not, as a consultant, client, or otherwise, pass on benchmarking findings to another organization without first getting the permission of your benchmarking partner and without first ensuring that the data is appropriately "blinded" and anonymous so that the participants' identity are protected.

7. **Principle of Completion**
   7.1. Follow through with each commitment made to your benchmarking partner in a timely manner.
   7.2. Complete a benchmarking effort to the satisfaction of all benchmarking partners as mutually agreed.
8. **Principle of Understanding and Agreement**
   8.1. Understand how your benchmarking partner would like to be treated, and treat him or her in that way.
   8.2. Agree how your partner expects you to use the information provided, and do not use it in any way that would break that agreement.
9. **Benchmarking with Competitors**
   The following guidelines apply to benchmarking with both actual and potential competitors:
   - In benchmarking with actual or potential competitors, ensure compliance with competition law. Always take legal advice before benchmarking contact with actual or potential competitors and throughout the benchmarking process. If uncomfortable, do not proceed. Alternatively, negotiate and sign a specific nondisclosure agreement that will satisfy the legal counsel representing each partner.
   - Do not ask competitors for sensitive data or cause the benchmarking partner to feel he or she must provide such data to keep the process going.
   - Do not ask competitors for data outside the agreed scope of the study.
   - Consider using an experienced and reputable third party to assemble and "blind" competitive data.
   - Any information obtained from a benchmarking partner should be treated as internal, privileged communication. If "confidential" or proprietary material is to be exchanged, then a specific agreement should be executed to specify the content of the material that needs to be protected, the duration of the

period of protection, the conditions for permitting access to the material, and the specific handling requirements that are necessary for that material.

Benchmarking Protocol

Benchmarkers

- Know and abide by the Benchmarking Code of Conduct for that part of the world.
- Have basic knowledge of benchmarking, and follow a benchmarking process.
- Prior to initiating contact with potential benchmarking partners, determine what to benchmark, identify key performance variables to study, recognize superior performing companies, and complete a rigorous self-assessment.
- Prepare a questionnaire and fully developed interview guide, and share these in advance, if requested.
- Possess the authority to share and be willing to share information with benchmarking partners.
- Work through a specified contact and mutually agreed-upon arrangements.

When the benchmarking process proceeds to a face-to-face site visit, the following behaviours are encouraged:

- Provide meeting agenda in advance.
- Be professional, honest, courteous, and prompt.
- Introduce all attendees, and explain why they are present.
- Adhere to the agenda.
- Use language that is universal; do not use jargon.
- Be sure that neither party is sharing proprietary or confidential information unless prior approval has been obtained by both parties, from the proper authority.
- Share information about your own process, and if asked, consider sharing study results.
- Offer to facilitate a future reciprocal visit.
- Conclude meetings and visits on schedule.
- Thank your benchmarking partner for sharing his/her process.

All employees of the organization who are engaged in any benchmarking activities usually are required to read and agree to follow this code of conduct.

The three keys to success are:

1. Have meaningful measurements and challenging targets.
2. Understand how and why others perform as they do.
3. Close all negative gaps.

Management's problems are:

- They don't know what level of performance is possible.
- They don't believe that major improvements are possible.
- They don't know how to bring about major breakthrough improvements within the organization.

Benchmark is the answer to all three of these problems.

Examples

> *A good benchmark is one that highlights an opportunity for improvement, not one that indicates you are the best.*
>
> **—H. James Harrington**

The following is a final outcome of a benchmarking study focusing on comparing the percentage of increase in employee salaries for a specific industry.

The figures that appear in the columns represent the 5th, 25th, 75th, and 95th percentile results achieved by businesses within the sample group size. The 50th percentile represents the median point of scores within the group.

BENCHMARK NAME: Staff Costs Growth (%)

Description: [(Staff costs – Staff costs previous year) / Staff costs previous year] × 100 [[(Q8 – Q8a) / Q8a] × 100]

This indicated the increase/decrease in staff costs of your business last year, compared with the previous-to-last year (see Figure 9.5 and Table 9.3).

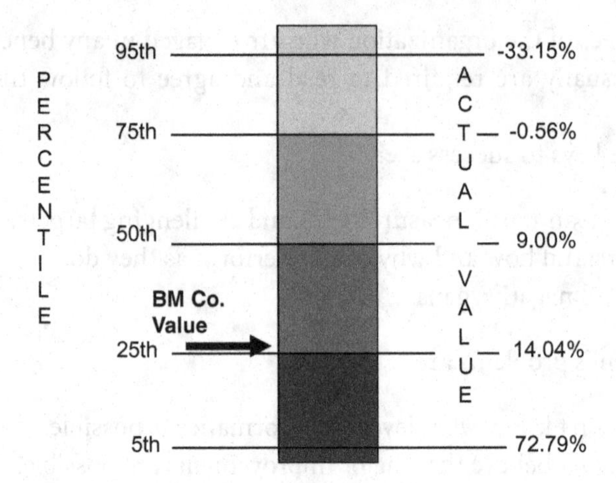

FIGURE 9.5
Comparison of Percent Increase in Staff Costs over One Year. *Source*: Centre for Organizational Excellence Research

TABLE 9.3

System Average Interruption Frequency Index

System Average Interruption Frequency Index (SAIFI)				
EDC	2006	2007	2008	2009
A-Company	0.14	0.25	0.26	
B-Company	1.16	.045	.046	
C-Company	0.79	0.68	0.67	
D-Company	0.79	0.79	0.99	
E-Company	1.35	0.99	1.04	
F-Corporation	1.27	1.11	1.05	
G-Company	1.50	1.63	1.07	
H-Company	1.22	1.19	1.13	
I-Company	1.16	1.29	1.16	
J-Company	1.73	1.63	1.35	
K-Company	1.47	1.71	1.56	
L-Company				2.78

The top-9 of the top-10 performing countries in order are:

• Trinidad	2.296
• South Korea	4.116
• Iceland	4.385
• Finland	4.394
• Germany	4.601
• Belgium	4.662
• Congo DR	5.242
• Netherlands	5.531
• Denmark	6.180

Software

Some commercial software available includes but is not limited to:

- MindMap www.novamind.com/
- Smartdraw www.smartdraw.com/
- QI macros http://www.qumacros.com

DESIGN FOR X

By Douglas Nelson and H. James Harrington – Harrington Management Systems

Design for X is the key to competitive, profitable products.

Introduction

Design for X (DFX) is both a philosophy and methodology that can help organizations change the way that they manage product development and

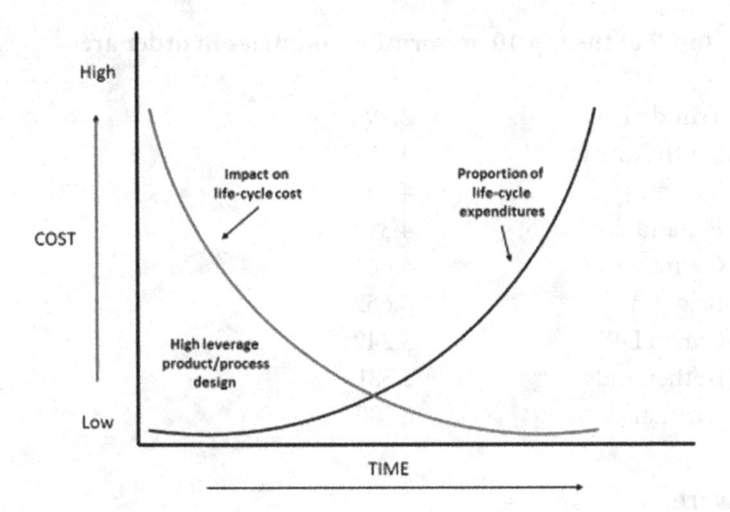

FIGURE 9.6
DFX Lifecycle

become more competitive. DFX is defined as a knowledge-based approach for designing products to have as many desirable characteristics as possible. The desirable characteristics include quality, reliability, serviceability, safety, user friendliness, etc. This approach goes beyond the traditional quality aspects of function, features, and appearance of the item.

AT&T Bell Laboratories coined the term "DFX" to describe the process of designing a product to meet the above characteristics. In doing so, the lifecycle cost of a product and the lowering of down-stream manufacturing costs was addressed (see Figures 9.6 and 9.7).

The Design for X makes use of the following 11 design methodologies:

1. Design for assembly
2. Design for servicing
3. Design for procurement
4. Design for manufacturability
5. Design for minimum cost
6. Design for reliability
7. Design for environmental
8. Design for reuse
9. Design for commonality
10. Design for disassembly
11. Design for safety

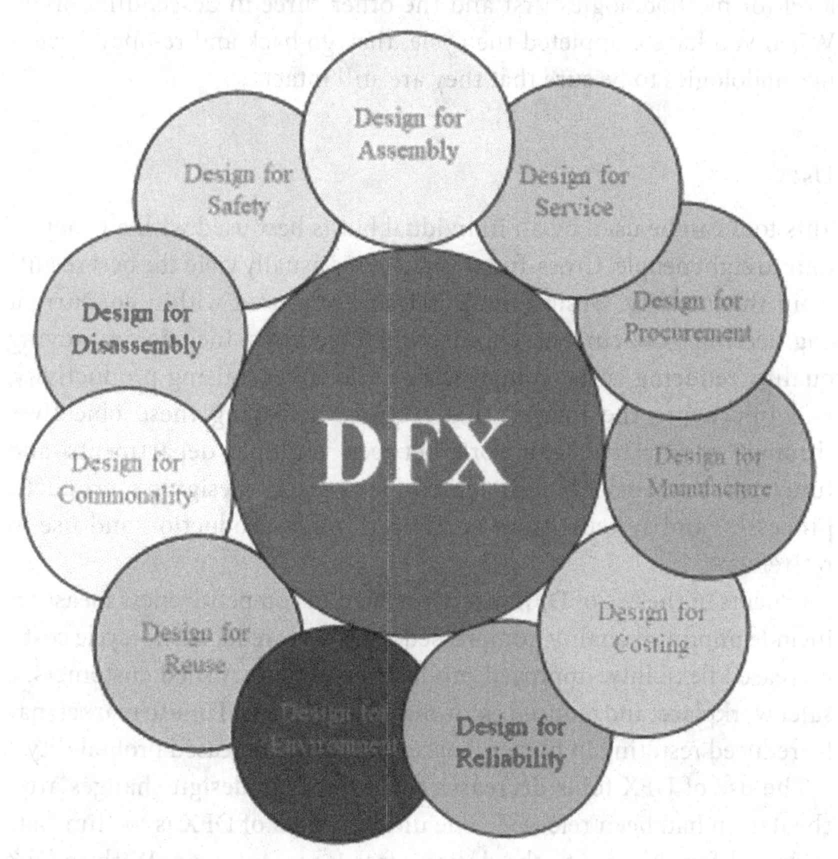

FIGURE 9.7
Design for X's Design Methodologies

To establish the order that the methodologies will be applied to the design, some companies classify each one of the 11 methodologies into one of the following categories:

- Level I – not critical
- Level II – important
- Level III – very important
- Level IV – most important

They will then start evaluating their design concepts applying the level for methodologies first and the other three in descending order. When you have completed the cycle, they go back and re-apply level I methodologies to be sure that they are still intact.

User

This tool can be used by an individual but is best used with a group of four to eight people. Cross-functional teams usually yield the best results from this activity. Design for X is frequently used within concurrent engineering. Concurrent engineering objectives include improving quality, reducing costs, compressing cycle times, raising productivity, and improving the image. This requires achieving these objectives through cooperative teamwork between multiple departments and functions to consider all interacting issues in designing products, processes, and systems from creation through production and use to retirement.

Benefits in the use of DFX directly related to competitiveness measures include improved quality, compressed cycle time, reduced life-cycle costs, increased flexibility, improved productivity, more satisfied customers, a safer workplace, and reduced environmental impacts. Time to market may be reduced resulting in higher market share and increased profitability.

The use of DFX tools decreases the number of design changes after the design had been released. The ultimate goal of DFX is to eliminate the need for changes to the design after it was released. With today's short product cycles, design problems that are discovered after we start shipping to customers often are not fixed due to the cycle time to make the change is greater than the product cycle time. And in addition to that, engineering errors detected by the user have an extremely negative impact on the customer's perception of the organization. There are also substantial savings because if the cost is X to make a change before the product is released to manufacturing, it is 10X after it is released and 100X after the product is shipped to an external customer.

Some additional benefits of applying DFX involve operational efficiency in product development. These may include:

1. Better communication and closer cooperation
2. Concurrence and transparency

3. Earlier customer and supplier involvement
4. Improved project management
5. Enhanced team environment
6. Better structuring of product development processes
7. Promotion of concurrent engineering

It is advisable to use DFX tools as early as possible in the design or redesign process. The Creation and Value Proposition Phase can provide valuable inputs to the DFX project. Outputs from the DFX project may be useful in refinement and updating within the Value Proposition Phase. As the innovative product and organization pass through the stages of Financing, Documentation, Production, and Sales/Delivery, significant cost reduction as well as quality and time improvements may be realized through early investment in DFX tools. These benefits will be borne out within the Performance Analysis stage of the Innovative Process.

How to Use the Tool

The earliest stages of product realization have the greatest effect on lifecycle costs, yet they represent the smallest proportion of lifecycle expenditures (based on Gattenby, David, et al.) (see Figure 9.6).

Any number of factors might be relevant to the definition of quality during a systems design initiative. DFX involves being able to incorporate a variety of factors "X" into a design, working toward a solution set that optimizes their interaction against customer needs and requirements. Common "X" factor examples include Assembly, Reliability, and Testing. Although this limited that thought pattern related to the engineering design that was acceptable in the 1980s, design concepts and requirements have increased significantly. The elegance of the design process output has expanded to include everything from creation through destruction.

The DFX toolbox has continued to grow in number from its inception to include hundreds of tools. Some researchers in DFX technology have developed sophisticated models and algorithms. The usual practice is to apply one DFX tool at a time.

DFX tools may be incorporated as part of a larger Concurrent Engineering, Product Development, Process Reengineering, Redesign,

Six Sigma, or TRIZ project. The tool is frequently used within concurrent engineering and can be used in new designs or in subsequent redesigns. The specific tool selected will depend on the product iteration and priority of goals and objectives relative to product design and cost and time constraints.

Knowledge Management System

One of the most valuable, if not the most valuable, tools in the Design for X toolbox is an effective Knowledge Management System (KMS). The KMS collects all the potential design possibilities in all 11 categories and stores it in its memory system to be used to evaluate and update each individual design. As time goes on, this database becomes more and more valuable as it collects potential error information that can be applied to the present design. To accomplish this, it has to contain information related to research and development, product engineering, manufacturing, field performance, customer misuse, and supplier internal and external related problems.

Design Guidelines

DFX methods are usually presented as design guidelines. These guidelines provide design rules and strategies. The design rule to increase assembly efficiency requires a reduction in parts count and part types. The strategy is to verify that each part is needed or that common part types could be used.

A number of general DFX guidelines have been established to achieve higher quality, lower cost, improved application of automation, and better maintainability. Examples of these guidelines for Design for Manufacturing include the following:

- Reduce the number of parts to minimize the opportunity for a defective part or an assembly error, allowing potential to decrease the total cost of fabricating and assembling the product, and to improve the chance to automate the process.
- Mistake-proof the assembly design so that the assembly process is unambiguous.

- Design verifiability into the product and its components to provide a natural test or inspection of the item.
- Avoid tight tolerances beyond the natural capability of the manufacturing processes and design in the middle of a part's tolerance range.
- Design "robustness" into products to compensate for uncertainty in the product's manufacturing, testing, and use.
- Design for parts orientation and handling to minimize non-value-added manual effort, to avoid ambiguity in orienting and merging parts, and to facilitate automation.
- Design for ease of assembly by utilizing simple patterns of movement and minimizing fastening steps.
- Utilize common parts and materials to facilitate design activities, to minimize the amount of inventory in the system and to standardize handling and assembly operations.
- Design modular products to facilitate assembly with building block components and sub-assemblies.
- Design considerations for serviceability into the product.

Design Analysis Tools

Each DFX tool involves some analytical procedure that measures the effectiveness of the selected tool. A typical design for assembly procedure should provide for analysis of such elements as handling time, insertion time, total assembly time, number of parts, and assembly efficiency. Each tool should have some method of verifying its efficiency.

While each design tool is typically considered individually, exploration of synergies and trade-offs can be useful in optimal positioning of products along the product lifecycle curve.

Design for X Procedure

Step 1: Product Analysis – Information related to the product is collected. Bills of Material (BOMs) are used to display product structure. Other types of product data can be easily correlated with the product BOM. It is useful to obtain the product hardware to examine and understand features.

Step 2: Process Analysis – Process Analysis is primarily concerned with the collection, processing, and reporting of process-specific and resource-specific data. Operation process and flow process charts are established.

Step 3: Measuring Performance – The process and product interactions can be measured in terms of the relevant performance indicators as relevant within the specified DFX tool. Additional data collection and processing may be required.

Step 4: Benchmarking – The objective is to determine whether or not the subject process is good and what areas contribute to it. Benchmarking primarily involves setting up standards and comparing the performance measurements against the established standard. Individual and aggregate benchmarks may be established. Once the performance standards and measurements are available, the areas where performance measurements are below standards can be identified.

Step 5: Diagnosis – Based on performance measurement and benchmarking, a determination is made as to what is good and what is not. In order to solve problems, it is necessary to know their causes. A cause-effect diagram may be useful in determining major causes for a problem. Root cause analysis techniques may be used to further identify specific conditions related to the problem.

Step 6: Advise on Change – Explore as many improvement areas as possible for each problem area. Brainstorming is a useful technique. Redesign of the product and processes is dependent upon specific circumstances. Changes may take place to composition, configuration, and characteristics at different levels of detail. Product changes may be made across the entire product ranges, working principles, concepts, structures, subassemblies, components, parts, features, or parameters. Process changes may be made across product lines, business processes, procedures, steps, tasks, activities, or parameters. Product and processes are closely interrelated. It is important to consider the interrelationships between the two when making a change to either products or processes. A "what-if" analysis may be helpful.

Step 7: Prioritize – The analysis may reveal a large number of problem areas within the product and processes. There may be many causes and solution alternates for each problem. Prioritization is often required due to limited resources. Prioritization should be based on some form of measurement. A Pareto chart may be constructed and can be used to show the relative frequency of events such as products, processes, failures, defects, and causes and effects. Analysis of the Pareto chart should aid in the prioritization of the problem areas.

Design for Product Lifecycle is a cradle-to-grave approach. Design for X methodologies can be used in design, production, useful-life, and finally, end of product life (See Figure 9.8).

There are currently hundreds of DFX tools that have been developed. A few of the more popular design tools are discussed herein.

Design for Safety

Design for safety requires the elimination of potential failure-prone elements that could occur in the operation and use of the product. The design should make the product safe for manufacturing, sale, use by the

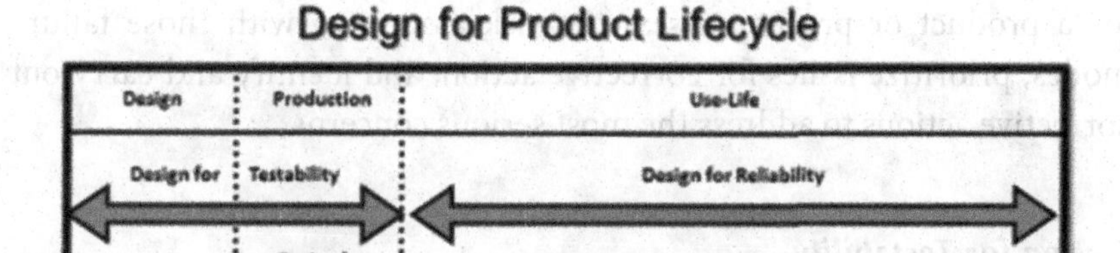

FIGURE 9.8
Design for Product Lifecycle

consumer, and disposal or reuse. Failure Modes and Effects Analysis (FMEA) and fault-tree analysis are often incorporated within Design for Safety. FMEA is a fundamental hazard identification and frequency analysis technique, which analyzes all the fault modes of a given equipment item for its (singular) effects on both other components and the system. A fault-tree analysis is a hazard identification and frequency analysis technique, which starts with the undesired event and determines all the ways in which it could occur. These are displayed graphically.

Design for Reliability

Design for Reliability describes the entire set of tools that support product and process design (typically from early in the concept stage all the way through to product obsolescence) to ensure that customer expectations for reliability are fully met throughout the life of the product with low overall lifecycle costs. A good tool to assess risk is the Failure Modes and Effects Analysis (FMEA). The FMEA is used to identify potential failure modes for a product or process, assess the risk associated with those failure modes, prioritize issues for corrective action, and identify and carry out corrective actions to address the most serious concerns.

Design for Testability

Design for Testability aims to make the product test procedures as easy and economical as possible during manufacturing, use, and servicing. Design for Testability includes techniques that add certain testability features to a hardware product design. The idea behind Design for Testability is that features are added to make it easier to develop and apply manufacturing tests for the designed hardware. The purpose of manufacturing tests is to validate that the product hardware is free of manufacturing defects that may affect the product's correct functioning. Tests are applied at several steps in the hardware manufacturing flow and may also be used for hardware maintenance in the field. The tests generally are driven by test programs that execute within Automatic Test Equipment (ATE) processes. These tests may also be conducted within the assembled equipment during maintenance procedures. In addition to finding and indicating the presence of defects, tests may log diagnostic information about the nature of the subject test fails. The diagnostic information can be used to locate the source of the failure.

Design for Assembly/Manufacturing

Design for assembly means simplifying the product so that fewer parts are required, making the product easier to assemble and the manufacturing process easier to manage. Design for Assembly is often the most effective DFX tool providing increases in quality while reducing costs and time to market. Design for Assembly is accomplished through the use of fewer parts, reduction of engineering documents, lowering of inventory levels, reduced inspection, minimization of setups, and materials handling. If a product contains fewer parts, it will take less time to assemble, reducing assembly costs. If the parts are provided with features that make them easier to grasp, move, orient, and insert, assembly time and assembly costs will be reduced. The reduction of the number of parts in an assembly has the additional benefit of generally reducing the total cost of parts in the assembly. Major cost benefits of the application of Design for Assembly are achieved through this reduction of the number of parts and, thus, cost of parts.

Design for Environment

Design for Environment aims to create minimal levels of pollution over the product lifecycle. Manufacture, use, and disposal are considered. The idea is to increase growth without increasing the amount of consumable resources. Some considerations include recovery and reuse, disassembly, waste minimization, energy use, material use, and environmental accident prevention. Design for Environment techniques may include lifecycle assessment, technology assessment, sustainable engineering, and sustainable design.

The Design for the Environment Program of the U.S. Environmental Protection Agency helps consumers, businesses, and institutional buyers identify cleaning and other products that perform well and are safer for human health and the environment. Lists are available that identify safer chemical products for use in manufacturing processes and also aid in determining safer reuse/recovery techniques or appropriate disposal.

Design for Serviceability

Design for Serviceability aims to return operation and use easily and quickly after failure. This is often associated with Maintainability. Design for Serviceability/Maintainability objectives include reduction of service requirements and frequency, facilitation of diagnosis, minimization of the time and costs to disassemble, repair/replace, and reassemble the product within the service process, and reduction of the cost of service components.

Design for Ergonomics

Human factors engineering must ensure that the product is designed for the human user. Some of the attributes that may be considered are: fitting the product to the user's attributes, simplifying the user's tasks (user-friendliness), making controls and functions obvious and easy to use, and anticipating human error.

Design specification best practices include (1) use of global anthropometry considerations (North America, Europe, Asia, and Latin America), (2) use of dimensions and ranges that support adjustability and reconfiguration, (3) designs to accommodate neutral postures and task variation, (4) minimization of manual material handling requirements, and (5) environmental considerations such as lighting, temperature, noise, and vibration.

Design for Aesthetics

Aesthetics is the human perception of beauty, including sight, sound, smell, touch, taste, and movement. Aesthetics is the aspect of design and technology which most closely relates to art and design, and issues of color, shape, texture, contrast, form, balance, cultural references; and emotional response are common to both areas. Products are becoming smaller and lighter. Customers desire that products be appealing in appearance. While considering customer image requirements, it is important to consider how the product will be manufactured to meet these aesthetic characteristics during the design process. Incorporation of a good industrial design leads to products that are genuinely appealing and represent a synthesis of form and function.

Design for Packaging

The most effective packaging for the product must be determined. The size and physical characteristics of the product are important. Design for automatic packaging methods may be considered. Packaging may be designed for maximum benefit in shipping or for product protection in distribution, storage, sale, and use. Design for packaging incorporates package design and development as an integral part of the new product development process. Products that are designed with packaging in mind can help ensure cost savings and product protection as it moves through the supply chain.

Design for Features

Design for Features considers the accessories, options, and attachments that may be used in conjunction with the product. Adding or deleting options is often used in creating products with similar manufacturing characteristics while meeting customer requirements within targeted marketing segments. It allows for the expansion of product line offerings without the cost and time involved in a complete redesign.

Design for Time to Market

Design for Time to Market helps to ensure that the timeliness of product launches may be maintained even as product lifecycles continue to shorten. The ability to make the product either earlier or faster than the competition can provide a market leadership advantage. Reducing time to market has a significant positive impact on revenue realization. With optimized processes, pre-launch development costs can be lower, time to launch can be faster, and market share gains can be faster and larger. An example of this is Toyota and their Prius automobile. By being first to market a hybrid gas/electric automobile, Toyota was in a position to establish market leadership in the hybrid market segment through first-mover advantage (see Liker, *The Toyota Way*). First-mover advantage is the advantage gained by the initial significant occupant of a market segment. Part of this advantage may be attributable to the fact that the first entrant can develop or gain control of resources that followers may not be able to match.

Summary

We here in the United States have earned the reputation of being the best problem solvers in the world, and that's probably because we've had more problems than anybody else to solve, so we got more experience. It's a little shocking to see that problem prevention is not being highlighted as a major requirement by ISO 9000. It looks like it is being replaced by risk management. I hope that with the focus on innovation and creativity, our emphasis will turn from correcting problems to designing problems out of our products and services. Certainly Design for X is one approach that goes a long way in making prevention a reality.

KNOWLEDGE MANAGEMENT EXCELLENCE

Today, more than ever before, knowledge is the key to organizational success. In order to fulfill this need, the Internet and other information technologies have provided all of us with more information than we can ever consume. Instead of having one or two sources of information, the Internet provides us with hundreds, if not thousands, of inputs, all of which need to be researched to be sure you have not missed a key nugget of information. We are overwhelmed with so much information that we don't have time to absorb it.

To make matters worse, most of the organization's knowledge is still not documented; it rests in the minds and experiences of the people doing the job. This knowledge disappears from the organization's knowledge base whenever an individual leaves an assignment. In *Knowledge Management Excellence – The Art of Excelling in Knowledge Management*, which is Book

IV in the series The Five Pillars of Organizational Excellence, we define how to establish a knowledge management system that will be designed to sort out unneeded or false information and capture the "soft" knowledge needed to run the organization.

With the almost endless amount of information that clouds up our computers, desks, and minds, a knowledge management system needs to be designed around the organization's key capabilities and competencies.

What Is Knowledge?

- *Definition of Knowledge:* Knowledge is defined as a mixture of experiences, practices, traditions, values, contextual information, expert insight, and a sound intuition that provides an environment and framework for evaluation and incorporating new experiences and information.
- *Definition of Knowledge Management:* Knowledge management is defined as a proactive, systematic process by which value is generated from intellectual or knowledge-based assets and disseminated to the stakeholders.

There are six phases required to implement an effective Knowledge Management System (KMS). These phases are:

- Phase I – Requirements Definition (7 activities)
- Phase II – Infrastructure Evaluation (16 activities)
- Phase III – Knowledge Management System Design and Development (12 activities)
- Phase IV – Pilot (15 activities)
- Phase V – Deployment (10 activities)
- Phase VI – Continuous Improvement (1 activity)

One of the biggest challenges related to implementing a KMS is transferring knowledge held by individuals, including processes and behavioral knowledge, into a consistent format that can be easily shared within the organization.

Knowledge takes us from chance to choice.

The true standard of success for knowledge management is the number of people who access and implement ideas from the knowledge networks. These networks bring state-of-the-art ideas or best practices into the workplace. This allows the organization to develop areas of critical mass that implement standards that work, and also provides access to everyone so they can make comments to improve those standards. Even the newest novice to the organization can look at the materials and make recommendations based upon personal insight, creativity, and experience.

A big challenge related to implementing a KMS is in transforming knowledge held by individuals, including process and behavioral knowledge, into a consistent technology format that can be easily shared with the organization's stakeholders. But the biggest challenge is changing the organization's culture from a knowledge-hoarding one to a knowledge-sharing culture.

Implementing a KMS

In a study conducted by *Knowledge Management* magazine, they asked the question, "What has been your biggest KM implementation challenge?" Here is what they discovered.

- 70% – Ensuring cooperation across business units
- 16% – Assembling the team
- 11% – Getting IS buy-in
- 2% – Staying on budget

Based on our experience, we would have rated changing the organization's culture as the number one problem.

> *Our first (KMS) initiative ignored the people element and paid limited attention to business readiness activities. We also lacked project sponsorship and effective communication – all the familiar potholes.*
>
> **—Sarah Dean, IS strategist, United**

Starting and Implementing a KMS

Knowledge management can help any business in four major areas: planning, customer service, training, and project collaboration. A knowledge management program should begin with the simple notion that to succeed it must be multifaceted. Looking at only one or two aspects can quickly lead to the self-defeating opinion that it simply won't work. This type of limited thinking leads to conclusions that may stall KM initiatives: Why manage knowledge if you can't measure it? What's the use of measuring it if management doesn't value it? Or why would anybody go to the trouble of sharing what they know?

Implementing successful KM solutions calls for an integrated approach and a multi-point action plan.[1] This requires an implementation approach that is essentially iterative, phase–based, and more widely inclusive than many traditional solution efforts.[2] Successful knowledge management solutions should be implemented using a model and process in an integrated approach. The Harrington Management Systems consulting organization has identified an approach, which is loosely based upon the Shewhart Cycle of Continuous Improvement model, that uses six key sets of implementation activities, which, when tied together with the proper infrastructure, can provide the foundation for successful KM solutions.[3]

The KMS implementation methodology we use considers seven fundamental things. They are:

- **Performance and Decision Modeling:** To define a knowledge strategy, one needs to clearly define the fundamental drivers of organizational performance and create a specific cause-and-effect relationship that links these drivers to company strategy. This activity enables an organization to link performance with effective process design and business rules decision-making.
- **Knowledge Enablers and Delivery Mechanisms:** The key factors that support the overall KM process and enable high performance are as follows:
  Information – the basic organized data required for the activity
  Knowledge bases – information organized in the actionable context
  Skill modules – the capability to act on the knowledge
  Collaboration – the ability to obtain help from others and provide it
  Action records – what has been done in the past
  Outcome data – the ability to relate past actions to results
  Shared insights – new ideas and recommendations that lead to even better results

- **Solution Design and Rapid Prototyping:** Once an organization has a clear design of what drives performance and the knowledge that enables that performance, solution design and rapid prototyping lay the groundwork for ultimate success. The goal is to iteratively build a prototype that integrates the various knowledge enablers into a performance support solution that is tied to the strategic performance model. A good solution design will clearly illustrate how performance will be improved.
- **Business Benefits Modeling:** Given a clear view of the performance goals of the organization and how the knowledge enablers in the solution are tied to the performance, a business benefits framework for the KM solution can be created. Business benefits for KM solutions can be quantified in a structured manner that links together the four perspectives of the Balanced Scorecard: Financial Well-Being, Customer Satisfaction, Internal Effectiveness, and Employee Creativity/Well-Being.
- **Content Management:** All successful KM solutions must create a process that continuously turns insights into knowledge that can be published and used by others in the organization. Implementing this process requires the definition and execution of several roles and responsibilities:
    - Infrastructure support supplies the necessary network-based tools and technologies to capture insights, edit structure, and broadcast to the larger user community.
    - Knowledge managers are responsible for establishing standards and templates for cross-organizational consistency and quality.
    - Contributors provide the fundamental content creation and editing capabilities using the standards from the knowledge managers.
    - Content owners possess in-depth domain expertise to ensure that knowledge is relevant, and they manage a team of contributors.
    - Librarians manage the knowledge library, ensure quality links between knowledge domains, and maintain the publishing support mechanisms.

While content management may seem to require significant additional effort, most organizations already have resources on board with these roles and responsibilities. What is typically missing is the alignment required to integrate these activities in a manner that collectively supports the strategic performance goals of the organization.

- **Change Management:** KM solutions often require significant change in personal and group culture and perception to achieve their full benefits. Although change management challenges are often daunting, the opportunity for success is significant. Often the act of implementing the KM solution is a change-management program itself.
- **Tying It All Together:** Using knowledge to drive competitive advantage is what knowledge management is all about. The costs to compete are going up, while the ability to maintain customers is going down. The new world of networked business relationships means that companies can no longer compete on scale, market share, costs, or other traditional metrics; instead, companies must compete on what they know.

KMS Maturity Grid

Before you start to install a KMS, we suggest you step back and think about what kind of "system" you have now, for every organization has some type of KMS in place. It may not be the most advanced system and it may not be working well, but it is working. No organization can exist without the ability to transfer knowledge to its employees in some way or another. One common way is through the internal "grapevine" inherent in every organization. However, the grapevine has a tendency to distort information, although it is fast, particularly if it is disseminating bad news.

Additionally, employees also get information from many other sources, i.e., from their managers and from outside sources. Some of the information we receive is not accurate, but we accept it as "fact" anyway. To take the first step, the organization needs to define the as-is status of its present KMS. We suggest using an independent organization to do an assessment of your KMS. However, the Twelve-Level Maturity Grid in Figure 9.9 is a quick way of ball parking the as-is status of your present KMS.

KEY CHANGE AREA: KNOWLEDGE MANAGEMENT

Scale

1. Knowledge management is not part of the organization's activities.
2. Knowledge management is part of the organization's vision, but no documented plan has been developed.
3. Knowledge management has a part of the three-year approved budget, but there is no implementation plan or measurement system in place.
4. Management is doing post-mortem of each project to define lessons learned, but there is no documented approach to be sure that everyone benefits from the analysis. A plan for implementing a full knowledge management has been prepared and is being implemented.
5. At the end of each project, lessons learned are defined, documented, and used to define how every future project will be implemented.
6. A central warehouse of best practices used in the organization's projects has been established and all employees have been trained on how to use them.
7. All clients are required to include best practice as defined in the data warehouse before a project is submitted for consideration. The approval of the project is contingent upon the inclusion of best practices.
8. A benchmarking clearing house has been established and data are collected from outside of the country.
9. The benchmarking clearing house includes international best practices and is used extensively throughout the organization.
10. Best practices are statistically correlated to their impact upon the countries' capacity improvement.
11. Knowledge is stratified to reflect how best practices differ based upon the area's development status.
12. The organization is recognized as the center of competency for knowledge management systems and is frequently best marked by other organizations outside of the organization.

FIGURE 9.9
The Knowledge Management Excellence Book

RISK MANAGEMENT

Introduction

Innovation and risk management are often thought of as two competing concepts. If innovation requires openness, trial and error, failure and a spirit of venturing out into the unknown, risk management, on the other hand, is thought of as a process of prevention and control that box things into what is known, safe, and certain.

The innovation standard published recently by ISO, ISO 56002:2019 Innovation Management System-Guidance clearly breaks this traditional view of risk and innovation. The standard integrates risk into innovation and associates risk with opportunities, thus encouraging decision-makers and innovators at taking a more proactive approach to managing innovation and risk. The standard mentions risk more than 30 times and makes it an integrated part of the innovation planning process. The goal is to help managers understand that the value of any innovation project is a function of risk and return, and encourage them to learn how to integrate risk into their thinking. When managers internalize risk management and adopt a proactive approach to identifying "what-if" scenarios, they are better equipped in handling the consequences of things going wrong.

Risk Management Framework

We usually define risk as the possibility of something going wrong. From this perspective, risk is not defined as an event, nor it is defined as an incident, but rather it is defined as a possibility of deviating from what you have originally planned. Global events, such as the 2008 financial crisis and COVID-19 pandemic, made the risk management discipline appear at the forefront of leadership discussion. Historically, risk management frameworks were better in mature banks and financial institutions. Other industries, where safety is critical, such as oil and gas, pharmaceuticals, and insurance, have also had better approaches to risk management. More recently, we see more organizations, even organizations in the public sectors and nonprofit sector, starting to integrate risk management into their thinking. The acceleration of change, the rapid shifts affecting the global economy, and the brutal disruptions produced by economic, social, and technological shocks during the past two decades are making more companies engage with the concept of risk. COVID-19, and its deep impact on the global economy, forced organizations to take a serious look at the need to establish a more comprehensive Enterprise Risk Management.

A Typology of Risks

We often divide risk into five categories. This typology comes from the practices of the financial sector as well as documents and international conventions regulating the banking industry such as Basel II.

- **Strategic risk**

  Strategic risk, or market risk, is the most important risk. It refers to the possibility of loss as a result of a change in the external environment of the organization. New threats from the competition, or changes in the economic environment, may adversely affect your position and will require you to adjust your strategy.
- **Operational risk**

  Basel II defines operational risk as the risk of loss resulting from a failure in internal process, people, and system. While Basel II addresses primarily the banking and financial industry, operational risk applies to all sort of organizations. Operational risk is very complex because it involves a variety of actors and a variety of processes.
- **Compliance risk**

  Compliance risk refers to the failure of an organization to comply with an industry standard or a government regulation. Compliance risk is important to address because it may affect the health and the well-being of people, as well as the reputation of the company when the organization is involved in lawsuits.
- **Financial risk**

  Financial risk involves the loss of value in an investment portfolio or a trading portfolio as a result of market change.
- **Security and system risk**

  A more recent topology of risk is emerging and relates to risk involving a breach in data and systems that could affect customer privacy, loss of data, and theft of intellectual property.

Ways of Dealing with Risk

The whole purpose of a risk management approach is to teach employees and managers on how to factor into their planning the possibility of something going wrong and raise the awareness level about the need to be prudent in day-to-day management. When risk management is systematized into a coherent framework, management is better prepared. It can decide to accept the risk, avoid it, transfer it, or design strategies to mitigate it. Let's take a look at these strategies.

- **Accept the risk**

  When you decide to ride your car on a snowy day rather than take a safer mode of transportation, such as a train, you have accepted the possibility of getting stranded, or even worse, getting into an accident as a result of the driving conditions. When you accept a risk, you are, in theory, aware of the consequences of your actions, and the probabilities of negative consequences occurring, and that you have decided to take full responsibility for it.

- **Avoid the risk**

  Risk avoidance is the opposite of risk acceptance. When you decide to stay home, rather than traveling, you have decided to avoid the risk of being stranded or getting into an accident. Similarly, when you exit a market, or you abandon an innovation project, you simply have decided to avoid the consequences of failing in the project.

- **Transfer the risk**

  In risk transfer, you make the decision to have a third-party deal with the consequences of the event. In this type of risk management strategies, you are aware of the event and the possibilities of it materializing, but you are asking a different party to manage the consequences. For instance, when you buy auto insurance, you are accepting the possibility of an accident happening, but you are transferring the consequences of the financial burden to the insurer.

- **Mitigate the risk**

  The goal of a mitigation risk approach is not to eliminate the risk but rather to manage it. In risk mitigation, you are accepting the risk, but you are proactively managing the negative consequences of the event, so you are not taken by surprise when the event actually occurs. In other words, mitigation simply means that you have thought about the risk and that you have taken some actions to limit its impact, like rescheduling a project, changing the team, or seeking for another partner.

Risk Management Framework: ISO 3100 Guidelines

A common mistake in thinking about risk management is to assign a person or a department to the function of risk and ask them to come up with a framework to identify, analyze, categorize, and mitigate risk and solutions. Because risk is everywhere, and risk tends to permeate all facets of the business, it is important that, when you think of innovation and risk, you take a comprehensive approach. ISO 3100:2018 Risk Management Guidelines provides a set of best practices that can be adapted by any organization of any size. The standard outlines eight principles that, if applied, should provide you with a great chance of coming up with a more comprehensive approach to establishing a risk management strategy.

1. **Risk integration**
   A risk management philosophy should be infused throughout the organization. When managers and employees are aware of the possibility of things going wrong while conducting their activities, you have better chances of managing the consequence of the event happening. Integration allows you to de-cluster the risk by making risk management tools and processes widely available throughout the organization.
2. **Structured and comprehensive**
   The second principle of risk management is to structure the risk management activity by a central hub for all activities related to risk management. The goal of the structure is to serve as a lead to risk management efforts and, at the same time, help create consistencies in the ways risk management is understood and applied. The structure also helps you create comprehensiveness in the way risk management processes are applied to different units.
3. **Customization**
   The third principle of risk management is customization. Customization requires an adjustment of risk tools and processes to fit the nature of the business or the unit where the tools are applied. A people approach to risk management will focus more on the behavior, for instance, while a cost approach will focus on market volatility impact on the project.
4. **Inclusive**
   Risk management framework must be inclusive of other partners and stakeholders' views and positions to develop a better perspective of what might affect the performance and the ability of these stakeholders to achieve their goals. A mature risk management framework helps manager connect with different stakeholders internally and externally. Inclusiveness means that you have thought about how to address the concerns of different players, which, in turn, should create more engagement, maximizing the chances of the project meeting its requirement. This ultimately will provide more transparency to stakeholders and partners on how you are managing the project.

5. **Dynamic**
   Because of the nature of markets and extreme shifts taking place in the world, the risk management approach needs to be able to adjust to changes and evolve as the external environment changes. Creating a risk management policy that does not get updated or ignores facts on the ground makes the risk management framework obsolete. Data aggregation and data integration, people risk, mindset, and behavior are all important trends that need to be addressed in your risk management approach.
6. **Best available information**
   When mapping out different risks in your organization, it is important that information you use as input is accurate, comprehensive, and part of what is best available. Building a scenario on outdated information or information that is not current, and does not lead to better understanding of risk, affects the credibility of the risk framework and diminishes the impact of mitigation strategies.
7. **The human and cultural factors**
   The seventh principle of the ISO guidelines 31000:2018 addresses the risk that relates to human behavior and the cultural mindset that runs in the organization. What we generally refer to as the "cultural risk" is the behaviors of people. Sometimes, frontline employees not trained on risk can be a source of trouble in data breaches, financial scandal, and legal exposure. A good risk manager should focus on behavior and train employees on how to identify early signs of risky behavior. Making risk identification as part of the cultural mindset of the organization provides the organization with an important layer of protection against threat.
8. **Risk improvement**
   The final principle suggested by ISO in building a risk management framework is the continual improvement of the formwork. When circumstances change and markets move, it is important for managers to update their risk analysis based on previous past performance of the tools and processes implemented.

The Risk Assessment Process

The Committee of Sponsoring Organizations of the Treadway Commission (COSO) provides a comprehensive model to risk identification and risk management. According to COSO, the purpose of Enterprise Risk Management is to identify the magnitude of risk in order to allow managers to prioritize their focus on the most critical risks. When assessing risks, organizations need to measure and identify the levels of priorities that need to get more or less attention, while defining the tolerance threshold that management can cope with. This, in turn, will help the organization define the contour of its risk appetite.

COSO model identifies six steps to an Enterprise Risk Management system.

1. **Identify risk**
   The first step in the COSO model is to identify the events, or the triggers, that might be of a risk. When conducting this activity, it is recommended that you categorize the events and their triggers. For instance, you can create a category for financial risk, people risk, strategic risk, or reputation risk depending on the nature of the innovation project you are running. A second recommendation, by COSO model, is to be as comprehensive as possible in identifying risks. Because this is a kind of a brainstorming activity about risks, it is helpful to identify as many as risks you can by casting the net wider.
2. **Develop assessment criteria**
   The second step in the process of mapping out risks is to develop criteria that would allow you to measure the level of risk. A common method used in assessing events is by looking at the likelihood of the event happening, and its impact in case it happens. For each event, you can ask yourself the following two questions:
   - On a scale of 1 to 5, what are the probabilities of the event happening? (Likelihood)
   - On a scale of 1 to 5, what are the impacts of the event when the event occurs? (Impact)
3. **Assessing the risk**
   In this step, a team of experts from different departments and business units who have knowledge about the business and the industry meet to conduct a qualitative and a quantitative analysis of the events identified in the previous step. The team should assign a value of 1–5 to each event, in terms of its probability to happen and its impact when it happens. The result is a quantitative analysis that maps out the risks and provides a visualization of risks (Figure 9.10).

Severity						
	5 Catastrophic	MEDIUM	MEDIUM	HIGH	EXTREM	EXTREM
	4 Major	LOW	MEDIUM	HIGH	HIGH	EXTREM
	3 Moderate	LOW	MEDIUM	MEDIUM	HIGH	HIGH
	2 Minor	LOW	LOW	MEDIUM	MEDIUM	MEDIUM
	1 Insignificant	LOW	LOW	LOW	LOW	MEDIUM
		Rare 1	Unlikely 2	Possible 3	Likely 4	Almost Certain 5
		Probability				

FIGURE 9.10
Risk Management Map

4. **Risk Interaction**
   Another step that helps to understand better the risk map is to address the risk connectedness and the outcome of risk interactions when two or three events collide. Because events don't happen in a vacuum, risk management takes a new dimension when two or three events interact with one another. For example, a liquidity problem can be the result of a weak economy. But when a liquidity problem in a weak economy interacts with a pandemic like COVID-19, the result may be totally different. A common approach to identify risk interaction is to use a boat-tie map, in which you identify vulnerabilities and their control to the left of the event, while identifying the impact of the mitigation and the remedy to the right of the event.
5. **Risk prioritization**
   Once all the risks are mapped out, you move to the prioritization of the risk of the events. Risks that get 5/5 on the probability dimension

and the impact dimension become your most critical risks. Senior management should focus on these events by proactively taking actions to mitigate them.

6. **Respond to risk**
The last step in the COSO model is to determine the type of actions you need to take for each event identified. You can either accept, transfer, avoid, or mitigate the risk, as we explained earlier.

Risk Management in ISO 56002:2019

The new innovation standard ISO 56002:2019 identifies a few areas in which risk is present in an innovation project. Below we will provide a discussion of these risks and provide some mitigation strategies.

- **Risk not identified during the planning phase:** During the planning phase of the innovation project, it is important that you set a time to discuss with your team the different types of risks that may affect, directly or indirectly, the project. Discussing openly the risks, and laying out the mitigation strategies that will help you to mitigate them will provide your team with a better approach to managing risks.
- **Uncertainties are not defined or factored into the project:** Changes that happen in the external environment may affect the innovation project in terms of funding, cost, and scheduling, which will have a direct impact on the completion and the launch of the product. It is important to run a variety of scenarios of "what-ifs" to set a response to the different variables and conditions that will have an impact on your project.
- **Lack of a system approach:** A system approach helps your team look at the risks from different angles. When you take a system approach, you are able to look at a variety of factors, such as human, cultural, legal, financial, process, and others, that may interfere with the completion of your innovation project.
- **Fear of failure and lack of risk taking:** While fear has a cultural dimension, it is important that you set the ambition of the innovation project at the right level. Innovation ambition should reflect what the organization can do, and what leadership aspires to achieve and sees as feasible. Stretching too much the capabilities of your innovation

team may lead to failure, which, in turn, creates a psychological barrier for future projects. It is also critical to teach your team how to get off the ground and learn from failure as much as they can, so they are able to fail fast and learn faster.

- **Portfolio approach:** A great way to manage risk in innovation projects is not to put all your eggs in one basket. The portfolio approach allows you to diversify your innovation projects, in terms of the degree of novelty and types of innovations, as well as time horizons. Spreading your innovation project over five years is a great way to balance the risk.
- **Budget constraints and lack of resources:** Funding innovation projects requires a commitment from the organization. It is critical that you integrate a risk management approach during the planning process so you are aware of the cost estimation. Carefully tracking the cost to forecast the evolution of the project and making the adjustment needed will help you to mitigate funding issues. Keep in mind that the most important cost to your organization is not to innovate.
- **Legal and regulations issues**: Compliance of the innovation project with industry standard and government regulation is critical, especially when the product is new. The risk on the user's health and well-being should be carefully studied and understood. Legal requirements must be also clarified to the team so there is no shortcut that may jeopardize the safety and may have legal consequences.
- **Deploying the innovation:** The risk of people not using or accepting the new product is real. Innovation is risky precisely because we don't know how people will react to it. So how you deploy the innovation and what logistics you have set for the deployment, including the marketing campaigns that will support the launch and the deployment, are of critical importance.

SURVEYS

We have selected surveys, one of the most popular and effective approaches to identify innovative opportunity, to provide you with the overall view of the survey's process. The following is some backup data related to conducting surveys in the United States.

- The average person is fed up with being asked to take surveys over the Internet or telephone. I average about four requests per week to contribute 5–10 minutes to take the survey each time. To get people to respond to surveys, the organization must be willing to pay the individual for their time. It is not unusual to be offered $10–$25 to fill out a specific survey.
- You cannot afford to have different functions within your organization, requesting surveys of your customer that have similar questions presented or that are not well-designed. As a result, many organizations are making marketing responsible for reviewing and approving all surveys that go to the organization's customers or potential customers.
- Frequently the person asking the questions has an accent that is difficult to understand because they are located in a country different from where they are surveying. Always match the survey with the country they are surveying.
- Give the individual an alternative asking, "Do you want to speak with the computer? Or do you want to speak with a live individual?"
- Don't collect any information you don't need or will not be using.

In our make-believe company, surveys are conducted in marketing, sales, research and development, product engineering, manufacturing engineering, manufacturing, production control, field services, strategic planning, and personnel. There are at least 10 different functions that are interested in obtaining feedback from the receiver of the organization's output. Surveys provide a means to upgrade the technical content and design questions that will provide the desired information. Please be sure that the information obtained reaches all parties interested in the subject matter and, even more important, the individuals who can use it to improve organizational performance and increase customer satisfaction. Great care should be taken in survey question design and data collection approaches. We recommend that an experienced single source be required to review and approve any customer/consumer surveys.

Conducting a survey is a very popular, low-cost way to collect information related to your organization from customers and potential customers. It frequently uses phone interviews with a random selection of individuals. They are usually conducted by organizations that have the wherewithal to make thousands of phone calls a day to random individuals throughout the target. The individuals making the telephone call have been trained in using a predefined sales pitch, but they are not confident to answer technical questions.

Designing Survey Questions

In deciding this survey, you need to consider things as given here:

1. What will the survey be used for?
2. Who will use the output?
3. How will the output data be analyzed?
4. What do I do to get people to respond to the survey?
5. Should it be a controlled sample or integrated into the regular activities?
6. Who will be sampled, and how big a sample size do you need?
7. What type of survey will you use?
8. How should the questions be designed?
9. For the multiple-choice questions, how many choices should I give them?
10. How do I distribute the surveys and collect the responses?
11. How do I to analyze the survey results?
12. What do I do with the analyzed results?

Next, you need to decide which questions you will ask and how you will ask them. It's important to consider the two major survey formats and how to develop a clear and precise question. There are two main forms of survey questions:

- Open-ended
- Closed-ended

Closed-ended questions give the respondent a predetermined set of answers to choose from. These questions are best for quantitative research.

They provide you with numerical data that can be statistically analyzed to find patterns, trends, and correlations.

Open-ended questions are best for qualitative research. This type of question has no predetermined answers to choose from. Instead, the respondent answers in their own words. Open questions are most common in interviews, but you can also use them in questionnaires. They are often useful as follow-up questions to ask for more detailed explanations of responses to the closed-ended questions.

You need to carefully consider each question in the survey. The question should be easily understood and difficult to misunderstand so the scope of the question will be understood:

- Sample 1: Is your manager participating in the way he/she gives you orders?
- Versus Sample 2: Is management participating in the way they deal with you?

All questions should be narrowly focused with enough context for the respondent to answer accurately. Design the survey so that participants have a way of indicating the question is not relative to them or that they have no experience related to the question so that they cannot answer. Always provide an opportunity for the person being surveyed to make personal comments in writing or verbally. Avoid questions that are not directly relevant to the survey's purpose.

When constructing closed-ended questions, ensure that the options cover all possibilities. If you include a list of options that isn't exhaustive, you can add an "other" field.

Readability is an important consideration in designing your questions. The people who are filling out the survey is a very important facet. The survey questions should be as clear and precise as possible. The questions should be tailored to your target population, keeping in mind their level of education and knowledge of the topic. If the survey is being conducted in a language that is a second language for the individuals participating, the questions should be worded accordingly.

Use language that respondents will easily understand and avoid words with vague or ambiguous meanings. Make sure your questions are phrased neutrally, with no bias towards one answer or another. Also, questions should be arranged in a logical order.

NOTES

1. This seven-point model is based upon the work of Renaissance Worldwide. Many of their solutions have been featured in *Knowledge Management* magazine over the past year or so and are cited as a practical model to the many problems that KM practitioners face when implementing their systems. They can be contacted at www.rens.com or call (781) 259-8833.
2. Ibid.
3. The implementation structure described is loosely based upon the writings of Renaissance Worldwide. They have described an approach by which knowledge management can fundamentally restructure an organization's competitive position. Understanding what drives performance and how it links to strategy is the first step. Identifying the knowledge that enables high performance, and rapidly creating innovative solutions to deliver that knowledge directly into the hands of the people that make decisions is imperative. Creating the content management capabilities and cultural environment that foster knowledge sharing and enable continuous improvement is the key to maintaining competitive advantage.

Bibliography

- COSO Risk Management in Practice 2012.
- Greenville Technical College book on Project Management Training (page 294).
- Harrington, H. J. and Lomax, Kenneth, (2000) Performance Improvement Methods, McGraw-Hill.
- Harrington, H. J. and Harrington, James S. (1996) High Performance Benchmarking – 20 Steps to Success, McGraw-Hill.
- Harrington, H. J. (1996) The Complete Benchmarking Implementation Guide – Total Benchmarking Management, McGraw-Hill.
- Harrington, H. J. (1998) The Creativity Toolkit—Provoking Creativity in Individuals and Organizations, McGraw-Hill.
- Harrington, H. J. (2006) Project Management Excellence: The Art of Excelling in Project Management, Paton Press.
- Harrington, H. J. and Trusko, Brett (2014) Maximizing Value Propositions to Increase Project Success Rates, Productivity Press.
- Harrington, H. J. and Harrington, James S. (1995) High Performance Benchmarking – 20 Steps to Success, McGraw-Hill.
- Huang, G. Q. (1996) Design for X, Concurrent Engineering Imperatives, Chapman and Hall.
- IEEE Std. 1149.1 (JTAG) Testability Primer, http://www.ti.com/lit/an/ssya002c/ssya002c.pdf.
- Reich, Robert B. (1992) The Work of Nations: Preparing Ourselves for the 21st, Century Capitalism, Vintage Books.
- Ritesh Jain, Fritz Nauck, Thomas Poppensieker, Olivia White, Managing Risk in the Age of Innovation.
- Scribbr B.V., website. located at 542 Singel, 1017AZ, Amsterdam, Netherlands.
- U.S. Environmental Protection Agency, http://www.epa.gov/dfe/.

Appendix A: The *Enhance Solutions Platform*™ for Innovation Management

The authors have selected the *Enhance Solutions Platform*™ to create a "live extension" to this book.

Dynamic process models representing many of the innovation process diagrams in the book have been created using *Enhance* and are available to you at no additional cost through *Enhance's TeamPortal*™ cloud service.

TeamPortal provides a way to dynamically navigate the book's process models, and it also supports additional knowledge assets such as documents, book extracts, and links to web-based resources and videos.

Combining this book with a cloud-based service like *Enhance* creates a unique and valuable way for you to learn how to implement and manage an innovative project or program within your enterprise. You can access *Enhance* by visiting **EDGESoftware.cloud/managing-innovation.**

We learn that creating individual process maps is really a simple but valuable activity. The complexity begins when you start weaving together many processes, considering process requirements, resource interconnections, and interdependencies in order to form a system usable for managing an innovation project. Then when you try to consider the Interfaces, dependency, and interrelationships between all of these systems, it becomes an extremely difficult matrix to understand, manage, and optimize. We find that to do the proper type of analysis, an advanced methodology implemented within an automated system is a valuable asset when you combine these many varying inputs and activities together. Just think of the number of tasks that make up an activity; then think of the number of activities that make up a process. Add to this the complexity of combining together 10 to 40 different processes, their interfaces, and interdependencies that make up a system. Finally, add the interconnections within systems when they try to connect them together considering all the way down to the task level.

This advanced methodology provides you with numerous benefits:

1. Drill Down into processes: See as much or little detail of your Knowledgebase as you need.
2. Interconnections: Access the same process model from many places in the Knowledgebase.
3. KnowledgeObjects: "Access the right knowledge at the right place at the right time."
4. Customize Best Practice processes: Change the Best Practice processes to fit the way you want to work.
5. Instant deployment of changes to everyone: No more worries about staff using out-of-date information.
6. Create a project and track its progress: Help manage an innovative project/program.

Here are some details to help you better understand each of the above items.

1. **Drill down into processes**

One of the challenges of navigating and understanding a large and complex Process Knowledgebase such as Managing Innovative Projects and Programs is not being overwhelmed by the large amount of information that is being presented. This "drill down" methodology allows you to start "at the top" of the Knowledgebase and select where you need more detail and then (double-click to) "drill-down" to see it. It also allows you to start at the top and "drill down" delivering a knowledge entity to the point where it needs to be used.

For example, here are some processes in the Managing Innovative Projects and Programs Knowledgebase that have been annotated with comments to help you understand this drill-down methodology. See Figure A.1.

By double-clicking on the embedded Process *PG 06. Resource Management*, you open up that Process model and access the next level of detail. What you access here is focused on just the *PG 06. Resource Management* Process. See Figure A.2.

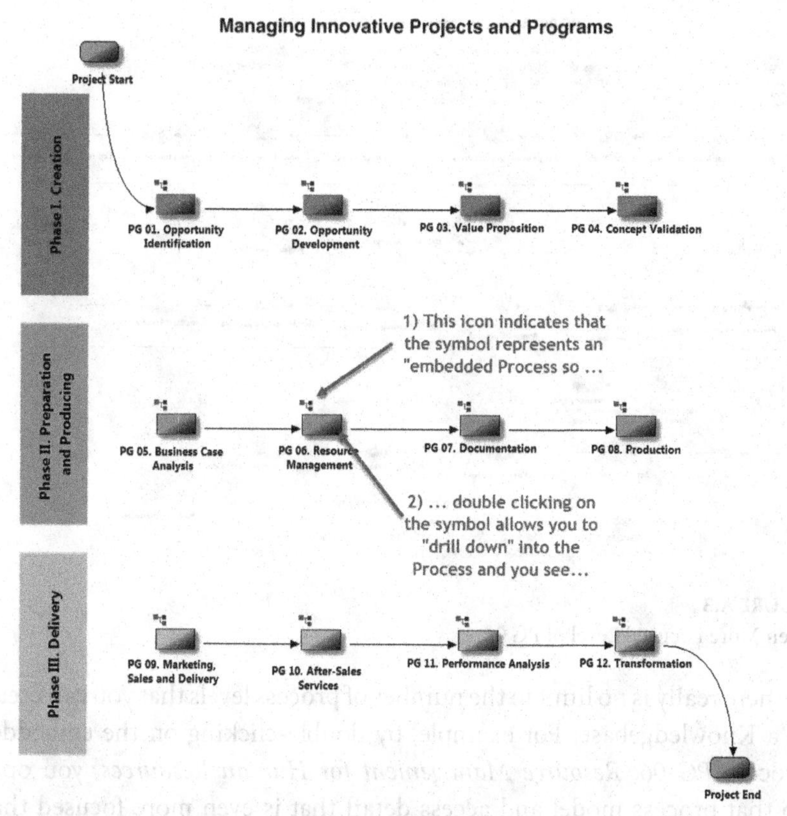

FIGURE A.1
Managing Innovative Projects and Programs Drill Down View

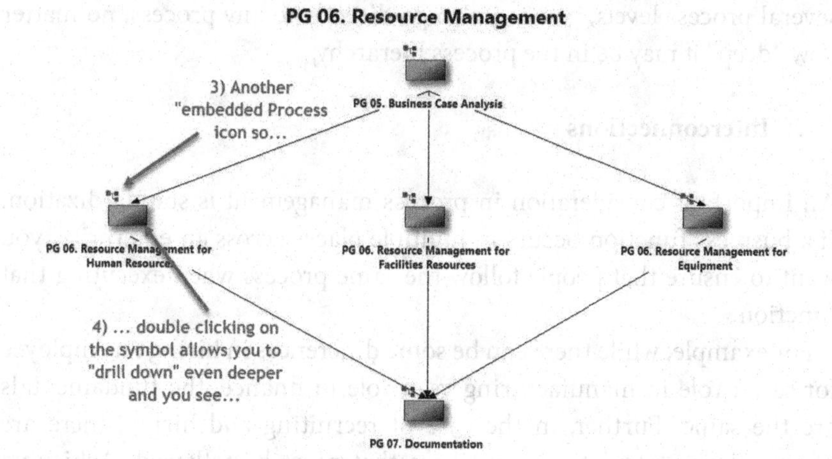

FIGURE A.2
Next Level of Details of Resource Management

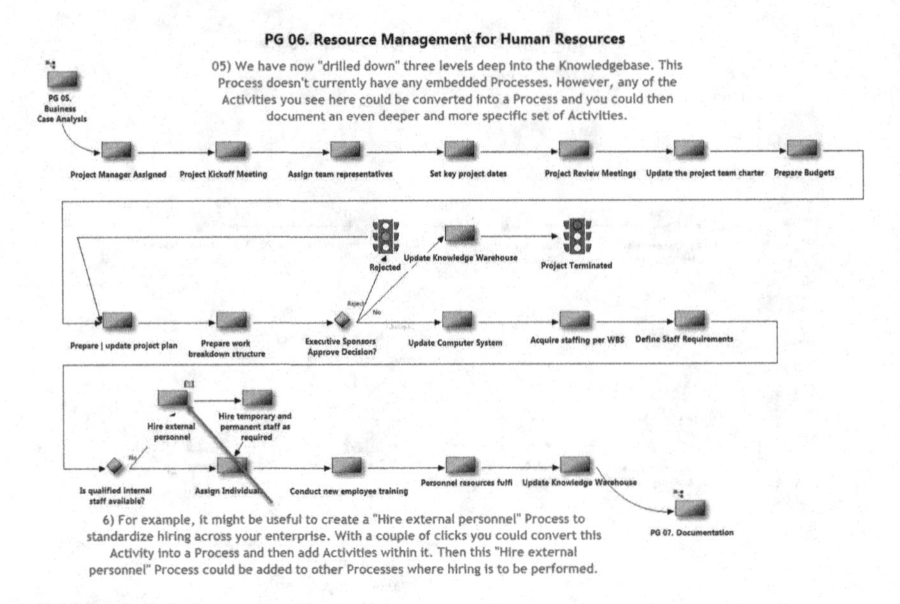

FIGURE A.3
Even More Focused Level of PG 06

There really is no limit to the number of process levels that you can create in a Knowledgebase. For example, by double-clicking on the embedded Process *PG 06. Resource Management for Human Resources*, you open up that process model and access detail that is even more focused than before. See Figure A.3.

In addition to starting at the top-level process and drilling down through several process levels, you can also go directly to any process, no matter how "deep" it may be in the process hierarchy.

2. Interconnections

An important consideration in process management is standardization. If a business function occurs in multiple places across an enterprise, you want to ensure that people follow the same process when executing that function.

For example, while there can be some differences in hiring an employee for say, a role in manufacturing vs. a role in finance, the fundamentals are the same. Further, in the case of recruiting and hiring, there are often policies, regulations, and laws that must be followed. Achieving these objectives can be accomplished by building a process model that captures all of this and linking it into all of the places in your Process Knowledgebase where appropriate.

In the prior section, you saw a sample that pointed out how the "Hire external personnel" activity could be converted into a process that would document your standard hiring process. Then that process could be placed in both manufacturing and finance processes that include a requirement to hire someone. From either of those processes, you can jump into the same hiring process, ensuring standardized execution of hiring in your enterprise.

As another example of the interconnectivity methodology, it is very common for an activity in one process to need to jump back into an earlier process. As an example, this can happen when an evaluation is made in one process with the outcome that the next activity should be to return to an earlier activity in another process to modify, say, a proposed design.

3. **KnowledgeObjects**

Access the right knowledge at the right place at the right time.

A Process model is essential in understanding and guiding how work should be done, but the value of the model can be significantly enhanced by attaching KnowledgeObjects to the activities in the process model.

A KnowledgeObject can be mostly any form of "digital knowledge." Common examples are documents, images, links to web pages, and even audio and video clips. For example, in a manufacturing environment, these knowledge assets might include:

- Job specifications
- Lists of necessary tools
- Required equipment
- Required fixtures
- Measurement equipment
- Measurement procedures
- Training plan

The value of attaching a KnowledgeObject to an activity is found in the fact that the process owner is best able to anticipate what additional knowledge a person will need to understand to perform an activity. By attaching that additional knowledge to the activity, we ensure that the person can "access the right knowledge at the right place at the right time"!

Here's an example from the Managing Innovative Projects and Programs Knowledgebase. See Figure A.4.

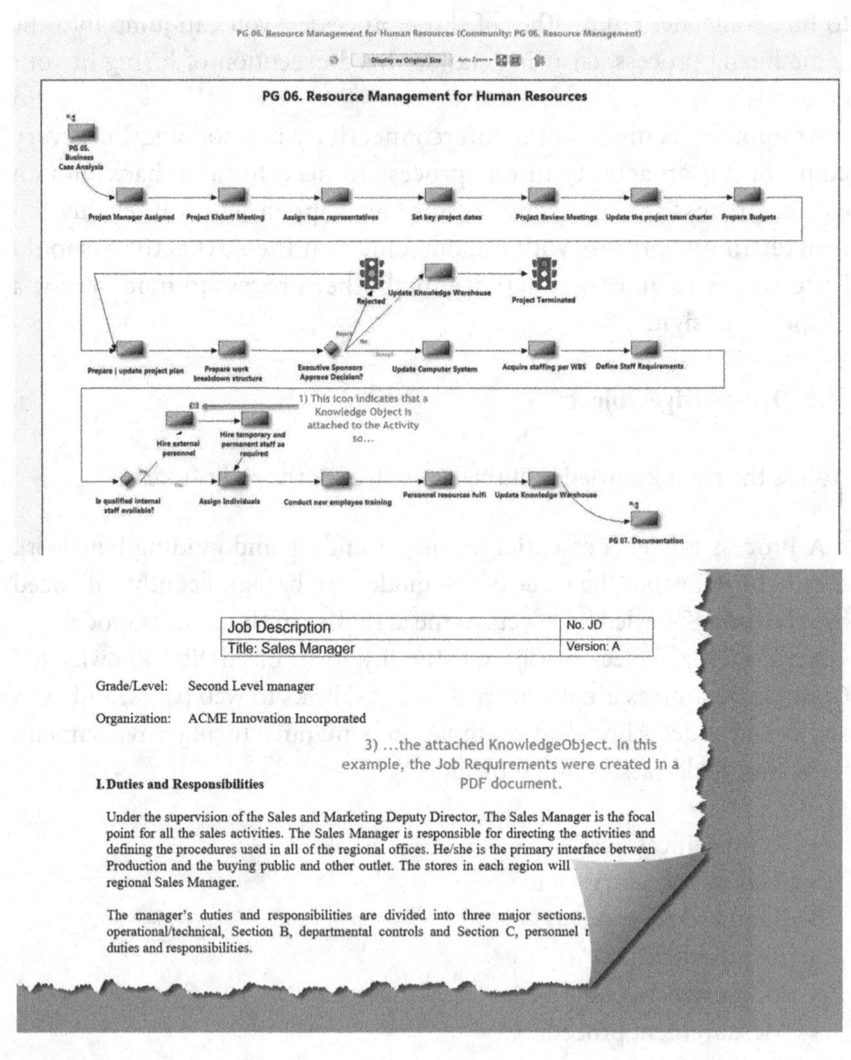

Job Description	No. JD
Title: Sales Manager	Version: A

Grade/Level: Second Level manager

Organization: ACME Innovation Incorporated

I. Duties and Responsibilities

Under the supervision of the Sales and Marketing Deputy Director, The Sales Manager is the focal point for all the sales activities. The Sales Manager is responsible for directing the activities and defining the procedures used in all of the regional offices. He/she is the primary interface between Production and the buying public and other outlet. The stores in each region will regional Sales Manager.

The manager's duties and responsibilities are divided into three major sections. operational/technical, Section B, departmental controls and Section C, personnel r duties and responsibilities.

FIGURE A.4
KnowledgeObject Added to Activity

4. **Customize best practice processes**

This methodology allows you to change Best Practice processes to fit the way you want to work. The authors of the book have used their innovation management expertise to create the Best Practice processes you have seen in the book. However, they realize that each enterprise is different, so the installed systems in an individual organization need to be customized to the unique customer, culture, objectives, and products that the organization deals with.

With *Enhance*, you can start with the Best Practice processes and easily modify them for the particular needs of your enterprise to create a Knowledgebase customized for your specific needs.

5. **Instant deployment of changes to everyone**

One of the reasons that process management is so important is that the world is changing every day. Your customers, competitors, vendors, the economy, and regulatory environment, to name just a few, are constantly changing. Your process models are a statement of "how we do work," and as a result, you have to keep your process models up-to-date. However, updated process models and KnowledgeObjects are of little value if people aren't aware of the changes.

Creating a Knowledgebase allows you to establish a "Single Point of Truth." One of the biggest failures in meeting performance objectives in an enterprise is caused by people using out-of-date instructions or other information. With the *Enhance* methodology, as soon as a process owner modifies a process model or KnowledgeObject, that updated knowledge is instantly available to everyone.

6. **Create a project and track its progress**

It is a challenge to manage a project whose processes require a large and complex Knowledgebase such as Managing Innovative Projects and Programs. With many of the processes being executed by multiple people at the same time, it can be very difficult to even know what has been accomplished and what is still to be done. Your management will certainly want to know!

Appendix B: Most Used Tools

LIST OF THE MOST USED AND MOST EFFECTIVE INNOVATIVE TOOLS AND METHODOLOGIES IN ALPHABETICAL ORDER

Book I – *Organizational and/or Operational IT&M*
Book II – *Evolutionary and/or Improvement IT&M*
Book III – *Creative IT&M*

Note: IT&M = Innovative Tools and/or Methodologies

P = Primary usage   S = Secondary usage   Blank = Not used or little used

	IT&M	Book III	Book II	Book I
1.	5 Why questions	S	P	S
2.	76 standard solutions	P	S	
3.	Absence thinking	P		
4.	Affinity diagram	S	P	S
5.	Agile Innovation	S		P
6.	Attribute listing	S	P	
7.	Benchmarking		S	P
8.	Biomimicry	P	S	
9.	Brain-writing 6-3-5-	S	P	S
10.	Business case development		S	P
11.	Business Plan	S	S	P
12.	Cause and Effect Diagrams		P	S
13.	Combination methods	P	S	
14.	Comparative analysis	S	S	P
15.	Competitive analysis	S	S	P
16.	Competitive shopping		S	P
17.	Concept tree (concept map)	P	S	
18.	Consumer co-creation	P		
19.	Contingency planning		S	P
20.	Co-Star	S	S	P
21.	Costs analysis	S	S	P
22.	Creative problem solving model	S	P	

	IT&M	Book III	Book II	Book I
23.	Creative thinking	P	S	
24.	Design for Tools		P	
25.	Directed/Focused/Structure Innovation	P	S	
26.	Elevator Speech	P	S	S
27.	Ethnography	P		
28.	Financial reporting	S	S	P
29.	Flowcharting		P	S
30.	Focus groups	S	S	P
31.	Force field analysis	S	P	
32.	Generic creativity tools	P	S	
33.	HU Diagrams	P		
34.	I-TRIZ	P		
35.	Identifying and Engaging Stakeholders	S	S	P
36.	Imaginary brainstorming	P	S	S
37.	Innovation Blueprint	P		S
38.	Innovation Master Plan	S	S	P
39.	Kano analysis	S	P	S
40.	Knowledge management systems	S	S	P
41.	Lead user analysis	P	S	
42.	Lotus Blossom	P	S	
43.	Market research and surveys	S		P
44.	Matrix diagram	P	S	
45.	Mind mapping	P	S	S
46.	Nominal group technique	S	P	
47.	Online innovation platforms	P	S	S
48.	Open innovation	P	S	S
49.	Organizational change mgt	S	S	P
50.	Outcome driven innovation	P		
51.	Plan-Do-Check-Act	S	P	
52.	Potential investor present	S		P
53.	Pro-active Creativity	P	S	S
54.	Project Management	S	S	P
55.	Proof of concepts	P	S	
56.	Quickscore creativity test –	P		
57.	Reengineering/Redesign		P	
58.	Reverse Engineering	S	P	
59.	Robust design	S	P	
60.	S-Curve Model		S	P

	IT&M	Book III	Book II	Book I
61.	Safeguarding Intellectual Properties			P
62.	Scamper	S	P	
63.	Scenario Analysis	P	S	
64.	Simulations	S	P	S
65.	Six thinking hats	S	P	S
66.	Social Networks	S	P	
67.	Solution Analysis Diagrams	S	P	
68.	Statistical Analysis	S	P	S
69.	Storyboarding	P	S	
70.	Systems thinking	S	S	P
71.	Synetics	P		
72.	Tree diagram	S	P	S
73.	TRIZ	P	S	
74.	Value analysis	S	P	S
75.	Value propositions	S		P
76.	Visioning	S	S	P
(P) Priority Rating	**Creative**	**Evolutionary**	**Organizational**	
TOTAL	29	23	24	

IT&M in Creativity Book 29
IT&M in Evolutionary Book 23
IT&M in Organizational Book 24

Index

A

AAA, *see* Area Activity Analysis
Activity block diagrams, 105, 108–109, 120
Aftersales service activities
activity block diagram, 217
inputs, knowledge management system, 217–220
complexity, 215
customer and supplier, 215
patience of Job, 215
phone calls, publisher, 216
representative, 216
sales/service book, 215–216
Agility, 31, 34
A-ha innovation, 8
Alphabetical order tools and methodologies, 303
Applied research, 71
Area Activity Analysis (AAA)
benefits of, 240–241
the cascading customer/supplier model, 240
definition, 239
features of, 241
five-level organization, 244
history of, 241–242
key 2020's tool, 238
methodology, 242
overview, 238–239
phase I, preparation for, 243
phase II, mission statement, 243–245
phase III, define area activities, 245
phase IV, develop customer relationships, 245–246
phase V, analyze activity's efficiency, 246
phase VI, develop supplier partnerships, 246–247
phase VII, performance improvement, 247–248
synopsis, 248
Artificial intelligence, 49, 51
Automatic Test Equipment (ATE), 270

B

Basic research, 71
Benchmark (BMK), 249
Benchmarking (BMKG), 60, 139
breakdown of, 253
code of conduct
agree principle, 257
competitors principle, 257–258
completion principle, 257
confidentiality principle, 256
contact principle, 255
exchange principle, 255–256
legality principle, 256–257
preparation principle, 255
understand principle, 257
use principle, 256
creation phase, 249
cycle time, 254
description, 255
evaluating competitive products, 254
examples, 259–261
ground rules, 252–253
improvement opportunity, 250
key performance indicators (KPIs), 250
performance analysis process, 249
performing countries, 261
processes types, 251
protocol
behaviours, 258–259
benchmarkers, 258
code of conduct, 259
software, 261
10 steps stairway success, 250–251
types comparisons, 251–252
types of, 250
value proposition process, 249
Blended approach, 164

BMK, *see* Benchmark
BMKG, *see* Benchmarking
Breakthrough innovation, 8
Business case analysis, 152–153
  activities, 160
  activity block diagram, 166–168, 170
  additional resources, 162–163
    criteria, rank business cases, 163–164
    results of, 164–165
  budget cycle, 161
  business case validation, 162
  document performance and project resource requirements, 162
  input(s), 160
  output(s), 160
  portfolio development team, 161
  synopsis, 169
Business case compliance, 163
Business process improvement methodologies, 239

C

Cascading customer/supplier model, 240
Clauses 1, 2, and 3, 15
Committee of Sponsoring Organizations of the Treadway Commission (COSO), 285–286
Competing for customers *vs.* growing market, 2
Complementarity, 32
Compliance risk, 282
Concept approval, 146–147
Concept validation, 147
  activity block diagram, 147–150
Context of organization, 24–25
Continuous improvement, 9, 151
Create, 5
Creation, 111
  Process Grouping 1 (*see* Opportunity identification)
  Process Grouping 2 (*see* Opportunity development)
  Process Grouping 3 (*see* Value proposition)
  Process Grouping 4 (*see* Concept validation)
  Tollgate I (*see* Opportunity analysis)
  Tollgate II (*see* Concept approval)
Creative, 5
  and innovative powers, 13–14
    10 Ss, 14–15
    Clauses 1, 2, and 3, 15
    knowledge management system (KMS), 16–17
Culture, 20, 34, 120
Customer-centric organizations, 199
Customer ship approval, 153, 199
  activity block diagram, 200–203

D

Data interpretation, 60
Data visualization, 60
Delivery, 207
  Process Grouping 9 (*see* Marketing, sales, and delivery)
  Process Grouping 10 (*see* After-sales service activities)
  Process Grouping 11 (*see* Performance analysis)
  Process Grouping 12 (*see* Transformation)
  return on investment, 151
  standard activities, 152
Department Activity Analysis, 238, 241; *see also* Area Activity Analysis (AAA)
Design for X (DFX)
  11 design methodologies, 262–263
  categories, 264–265
  definition, 262
  design analysis tools, 267
  design guidelines, 266–267
  knowledge management system (KMS), 266
  lifecycle, 262
  philosophy and methodology, 261
  procedure, 267–269
    aesthetics design, 272
    assembly/manufacturing design, 271
    environment design, 271
    ergonomics design, 272
    features design, 273

packaging design, 273
reliability design, 270
safety design, 269–270
serviceability design, 272
testability design, 270–271
time market design, 273
synopsis, 274
tool use, 265–266
user of
benefits, 264
concurrent engineering, 264
operational efficiency, 264–265
project, 265
tools, 265–266
Developed research, 71
DFX, *see* Design for X
DFX lifecycle, 262
Discretionary financial resources, 179
Documentation, 153
activity block diagram
marketing and sales documentation, 192–195
producing output controls documentation, 190–192
production set up documentation, 188–190
product specifications documentation, 183–184
project management plan and facilities planning, 185–186
suppliers and contractors documentation, 187–188
document control and management, 181
information, 181
Innovative New Entity Cycle, 182
six document management systems, 182
synopsis, 195

E

Effective Portfolio Management Systems, 154
Empowerment, 215
Enhance Public Knowledgebase, 99
Enhance Solutions Platform™
best practice processes, 301
deployment changes, 301
drill down processes, 296–298
interconnections, 298–299
knowledge objects, 299–300
methodology, 295
project creation and track progress, 301
TeamPortal, 295
Entity, 39
Equipment resource management, 176–177
Evolutionary innovation, 9
External assessor, 58

F

Failure Modes and Effects Analysis (FMEA), 270
Financial risk, 282
Five Levels of Innovation, 9
Five tollgates, 96–97
Floor space resource management, 174–176
Flowcharting, 101, 105

G

Gradual innovation, 9

H

Hard drivers, 15
Hiring process flowchart, 106
Horizontal activity block diagram, 108–109
Hybrid, 39

I

Ideal improvement process, 225
IMA, *see* Innovation Management Assessment
Improvement, 29
types, 9–10
apparent solutions, 10
discoveries, 11

major improvements, 10
minor improvements, 10
new paradigms, 10–11
IMS, *see* Innovation Management System
Individual Performance Indicators (IPI), 238
Initiative, 37
Cambridge, 37
Webster, 37
Innovation, 5–7, 31
types, 8, 11
opportunities, 119–120
subcategories, 8–11
Innovation Management Assessment (IMA), 55
Innovation Management System (IMS), 16, 19, 53, 65, 102
advantages, 51–53
maintaining advancements, 52
upgrading, 53
Innovation Project Team (IPT), 83, 126
Innovation strategy, 62
components, 62
activities, 62
communication, 62–63
documentation, 63
flexibility, 62
Innovative, 39
designer, 48
organizational structure, 39
thinking
discouraging phrases, 12
innovative power techniques, 12–13
Innovative New Entity Cycle, 182
Internal assessor, 58
International Association of Innovation Professionals (IAOIP), 237
International Organization for Standardization (ISO), 19
IPT, *see* Innovation Project Team
ISO, *see* International Organization for Standardization
ISO 56000 innovation management, 63
ISO 56000:2020 Standards, 21
ISO 56002:2019, xv
ANSI, xvii–xviii
budget constraints and lack resources, 289
deploying innovation, 289
fear of, failure and lack, 288–289
innovation management, 63
Innovation Management Systems Standard
hierarchy understanding, xvi–xvii
innovation processes cycle, 89–90
legal and regulations issues, 289
organizational structure
clause 4.0, context of organization, 24–25
clause 5.0, leadership, 25–26
clause 6.0, planning, 26–27
clause 7.0, support, 27–28
clause 8.0, operations, 28
clause 9.0, performance evaluation, 28–29
clause 10.0, improvement, 29
planning phase, 288
portfolio approach, 289
Standards, 20, 22–24
10 clauses, 23–24
TIME methodology, 20–21
system approach, 288
uncertainties, 288
ISO's Technical Committee 279, 21
ISO TR 56004:2019 Innovation management assessment, 63

K

Key business considerations, 163
KMS, *see* Knowledge management system
Knowledge, 275
Knowledge management scale, 280
Knowledge management system (KMS), 16–17, 266
challenge, 276
implementation challenge, 276
implementation methodology
business benefits modeling, 278
content management, 278–279
knowledge enablers and delivery mechanisms, 277
performance and decision modeling, 277
solution design and rapid prototyping, 278

information, 274–275
maturity grid, 279
networks, 276
phases, 275
starting and implementing, 276–277

L

Leadership, 25–26, 60, 62

M

Management innovation, 8
Manufacturing engineering, 73–74
Marketing, 74–75
typical activities
distribution, 75
financing, 76
forecasting, 76
pricing, 75–76
product/service management, 75
promotion, 75
selling, 75
Marketing, sales and delivery
activity block diagram, 211–214
competition, 208
consistent strategies, 210
customer service, 209
distribution, 208
financing, 209–210
five Cs of marketing, 208
five Ss of sales, 210
markets, 208
positioning, 209
pricing, 209
promotion, 209
seven sales items, 210–211
supply chain, 208
synopsis, 214–215
McKinsey 7S model, 14
"Must comply" criteria, 164

N

Natural work teams (NWTs), 238, 240–241

O

Operational control, 30
Operational risk, 282
Operations, 28, 30
clause 8.1, operational planning and control, 30–31
collaboration and partnership, 31–36
clause 8.2, managing innovation initiatives, 36–37
subclause 8.2.1, managing each innovative initiatives, 37–39
subclause 8.2.2, implementation of innovative initiative, 39–41
clause 8.3, innovation processes, 41
subclause 8.3.1, IMS overview, 41–43
Opportunity analysis, 123–125
Opportunity development, 125–126
activity block diagram
for apparent or minor opportunities, 129–130
for major opportunities, 130–132
for new paradigms and discovery opportunities, 132–135
objective, 126–129
Opportunity identification, 111–115
activity block diagram, 119–123
inputs, 115–119
Opportunity selection box, 128
Optimization, 42
Organizational agility, 34–35
Organizational structure, 34, 39

P

PAR's, *see* Project Accomplishment Reviews
Participative Management, 83
Performance analysis, 220–221
activity block diagram, 225–228
synopsis, 228
Performance evaluation, 28–29
Performance of, process problems
lack of employees' engagement, 36
lack of employees' skills, 36
lack of governance, 35

PIC, *see* Project Innovation Cycle
PIC cast members
key players, 65–66
team ground rules, 84–87
typical business process list
executive project sponsor, 76–77
innovation project team, 83
manufacturing engineering, 73–74
marketing, 74–76
product engineering, 72–73
project managers, 78–83
project team leaders, 77–78
research and development, 70–72
team leaders, 78
typical business process list, 66–70
Planning, 26–27
PMBOK methodologies, 46, 138
Preparation and producing, 151
continuous improvement, 151
creative/innovative cycle, 152
Process Grouping 5 (*see* Business case analysis)
Process Grouping 6 (*see* Resource management)
Process Grouping 7 (*see* Documentation)
Process Grouping 8 (*see* Production)
Tollgate III (*see* Project approval)
Tollgate IV (*see* Customer ship approval)
Process activity block diagram, 121
Process documentation levels interaction, 101
Process innovation, 8, 150
decision-making, 34
improvement, 121
opportunities, 120
Product activity block diagram, 120
Product engineering, 72–73
Product innovation, 8
opportunities, 119
system assessment, 65
Production, 153, 195–196
activity block diagram, 196–199
synopsis, 203–205
Product lifecycle design, 269
Project Accomplishment Reviews (PAR's), 95
Project approval
activity block diagram, 154–158
business case analysis, 158–159
synopsis, 169
Project evaluation, 221
improvement methodology results, 221–222
top five positive/negative innovation change impacts, 222–225
Project innovation, 1–3
addressing, unaddressable, 4–7
creative and innovative powers, 13–14
10 Ss, 14–15
Clauses 1, 2, and 3, 15
knowledge management system (KMS), 16–17
factors affecting innovation, 7–8
innovation killers, 11–13
innovation types, 8, 11
subcategories, 8–11
Project Innovation Cycle (PIC)
phases, 90–92
phase I, 111–150 (*see also* Creation)
phase II, 151–205 (*see also* Preparation and producing)
phase III, 207–235 (*see also* Delivery)
Tollgates, 93–96
Project managers, 78–80
leadership attributes, 80–81
project deliverables, 81–82
weekly checklist, 82–83
Project sponsor, 76–77
responsibilities, 77
Project team leaders, 77–78

Q

Qualitative/quantitative criteria, 164
Quality Circles approaches, 239

R

R&D department, 70
Resource management, 153
activity block diagram, 170
facilities resource management, 174–177

facilities setup resource management, 177–179
financial resources management, 179–181
human resource staffing, 171–174
Innovative Project Team (IPT), 170
inputs process, 170–171
organization's portfolio resources, 169
Risk management
assessment process, 285–288
assessing, 286
development criteria, 286
identify risk, 286
interaction, 287
prioritization, 287–288
respond, 288
dealing, 282
accept, 282–283
avoid, 283
mitigate, 283
transfer, 283
framework, 281
ISO 3100 Guidelines, 283–285
typology risks, 281–282
innovation, 280–281
ISO 56002:2019 identifies
budget constraints and lack resources, 289
deploying innovation, 289
fear of, failure and lack, 288–289
legal and regulations issues, 289
planning phase, 288
portfolio approach, 289
system approach, 288
uncertainties, 288
Risk management map, 287

S

SAIFI, *see* System Average Interruption Frequency Index
Sales and marketing innovation, 8
Security and system risk, 282
Service Activity Block Diagram, 121
Service innovation, 8
Service innovation opportunities, 120
"Shall," 23
"Should," 23
Six levels of, documentation, 99–101
activities divided into tasks, 107–108
organizational, 102
phases divided into Process Groupings, 103–104
Process Groupings divided into processes, 104
process levels divided into activities, 105–107
systems, 102
Six Sigma system, 83
Soft drivers, 15
Staff costs, percent comparison, 260
Stairway to success, 251
Standard 56004:2019, 55
assessment approaches, 55
assessment focus, 57
assessment objectives, 56–57
breath and extent of assessment, 57
data collection and tools, 58–59
data interpretation, 60
data type, 59
expertise involved in assessment, 58
IMA checklist, 61
reference and comparison, 59–60
innovation management
outputs, formats, and reports, 60
performance criteria, 62–64
Strategic risk, 282
Strategy, 62
Structures, 39
Suboptimization, 42
Support, 27–28
Surveys
closed-ended questions, 292
collect information, 291
designing survey questions, 291–292
functions, 290
language, 292
open-ended questions, 292
readability, 292
scope of, 292
United States, 290
"Sustaining Sponsor," 164
System Average Interruption Frequency Index (SAIFI), 260

T

Team ground rules, 84–87
Team leaders, 78
Team member responsibilities, 83
10 Ss, 14–15
The Art of War, 251
TIME methodology, 20–21
TIME Pyramids for innovation, 43, 44
    building blocks of, 43
        commitment to stakeholders' expectations, 45
        comprehensive measurement systems, 50
        individual creativity, innovation, and excellence, 47
        innovative design, 47–48
        innovative executive leadership, 45
        innovative management participation, 46
        innovative organizational assessment, 44
        innovative organizational structure, 50
        innovative project management systems, 46
        innovative robotics/artificial intelligence, 48–49
        innovative supply chain management, 47
        innovative team development, 46
        knowledge assets management, 49–50
        performance and cultural change management plan, 45
        rewards and recognition process, 50–51
        value to stakeholders (the foundation), 43
Tollgate, 93; *see also* Tollgate *specific entries*
Total Quality Control (TQC), 239
Total Quality Management (TQM), 225, 239
Transformation
    activity block diagram, 231–235
    analysis, 230–231
    commitment document, 229
    customer's representatives, 230
    cycle time reductions, 230
    evaluation focus, 230
    innovative measurements, 229
    projected estimates, 229
    synopsis, 235
Typical ground rules
    for SS team, 87
    for team leaders, 85–86
Typology risks, 281–282

U

Unique, 4–5, 114–115

V

Value added, 3–4, 128–129
Value proposition, 135, 138
    inputs, 136–142
    and Tollgate II, activity block diagram, 142–145
Value Proposition Development and *Business Case Development*, 161
Vertical activity block diagram, 109
Visualization, 60, 286

X

X techniques, design for, 48

Printed in the United States
by Baker & Taylor Publisher Services